Mathematical Gnostics

The book describes the theoretical principles of nonstatistical methods of data analysis but without going deep into complex mathematics. The emphasis is laid on presentation of solved examples of real data either from authors' laboratories or from open literature. The examples cover wide range of applications such as quality assurance and quality control, critical analysis of experimental data, comparison of data samples from various sources, robust linear regression as well as various tasks from financial analysis. The examples are useful primarily for chemical engineers including analytical/quality laboratories in industry, designers of chemical and biological processes.

Features:

- Exclusive title on Mathematical Gnostics with multidisciplinary applications, and specific focus on chemical engineering.
- Clarifies the role of data space metrics including the right way of aggregation of uncertain data.
- Brings a new look on the data probability, information, entropy and thermodynamics of data uncertainty.
- Enables design of probability distributions for all real data samples including smaller ones.
- Includes data for examples with solutions with exercises in R or Python.

The book is aimed for Senior Undergraduate Students, Researchers, and Professionals in Chemical/Process Engineering, Engineering Physics, Stats, Mathematics, Materials, Geotechnical, Civil Engineering, Mining, Sales, Marketing and Service, and Finance.

Mathematical Gnostics
Advanced Data Analysis for Research and Engineering Practice

Pavel Kovanic

CRC Press is an imprint of the
Taylor & Francis Group, an **informa** business

First edition published 2023
by CRC Press
6000 Broken Sound Parkway NW, Suite 300, Boca Raton, FL 33487-2742

and by CRC Press
4 Park Square, Milton Park, Abingdon, Oxon, OX14 4RN

CRC Press is an imprint of Taylor & Francis Group, LLC

© 2023 Taylor & Francis Group, LLC

Reasonable efforts have been made to publish reliable data and information, but the author and publisher cannot assume responsibility for the validity of all materials or the consequences of their use. The authors and publishers have attempted to trace the copyright holders of all material reproduced in this publication and apologize to copyright holders if permission to publish in this form has not been obtained. If any copyright material has not been acknowledged please write and let us know so we may rectify in any future reprint.

Except as permitted under U.S. Copyright Law, no part of this book may be reprinted, reproduced, transmitted, or utilized in any form by any electronic, mechanical, or other means, now known or hereafter invented, including photocopying, microfilming, and recording, or in any information storage or retrieval system, without written permission from the publishers.

For permission to photocopy or use material electronically from this work, access www.copyright.com or contact the Copyright Clearance Center, Inc. (CCC), 222 Rosewood Drive, Danvers, MA 01923, 978-750-8400. For works that are not available on CCC please contact mpkbookspermissions@tandf.co.uk

Trademark notice: Product or corporate names may be trademarks or registered trademarks and are used only for identification and explanation without intent to infringe.

ISBN: 978-1-138-33923-1 (hbk)
ISBN: 978-1-032-42351-7 (pbk)
ISBN: 978-0-429-44119-6 (ebk)

DOI: 10.1201/9780429441196

Typeset in Latin Modern
by KnowledgeWorks Global Ltd.

Contents

Preface	xi
Introduction	xv
Author Biography	xix

1 Introductory Kindergarten   1
   1.1 Elemental Notions 1
       1.1.1 Abelian Group 2
       1.1.2 Variability 2
       1.1.3 Morphism and Invariant 3
       1.1.4 Vector Space 4
       1.1.5 Matrices 4
       1.1.6 Probability Distribution 6
   1.2 Sources of Inspiration for Mathematical Gnostics 6
       1.2.1 Theory of General Systems 7
       1.2.2 Theory of Measurement 7
       1.2.3 Geometries 7
       1.2.4 Maxwell's Contributions 8
       1.2.5 Relativistic Physics 8
       1.2.6 Thermodynamics 8
       1.2.7 Matrix Algebra 8
   1.3 Conclusions 9

2 Axioms   11
   2.1 Axioms of the Data Model 11
   2.2 Applications of Axiom 1 12
   2.3 Data Aggregation as the Second Gnostic Axiom 14
   2.4 Conclusions 14

3 Introduction to Non-Standard Thought   15
   3.1 Paradigm 15
   3.2 Statistical Paradigms 17
   3.3 Statistical Data Weighing 18
   3.4 Non-Statistical Paradigms of Uncertainty 19
   3.5 On the Need of an Alternative to Statistics 20
   3.6 Principles of Advanced Data Analysis 23

3.7	The Gnostic Concept	25
3.8	Conclusions	27

4 Quantification   29
4.1	Ideal Quantification	29
4.2	Real Quantification	31
4.3	Conclusions	33

5 Estimation and Ideal Gnostic Cycle   35
5.1	A Game with Nature	35
5.2	Double Numbers	36
5.3	Gnostic Data Characteristics	36
5.4	The Ideal Gnostic Cycle	39
5.5	Information Perpetuum Mobile?	40
5.6	Existence and Uniqueness of the Ideal Gnostic Cycle	40
5.7	Conclusions	41

6 Geometry   43
6.1	A Historical Dispute on Robustness of Statistics	43
6.2	Distance as a Problem	45
6.3	Additivity in Data Aggregation	47
	6.3.1 Statistical Mean Value and Data Weighting	48
6.4	Double Robustness	49
6.5	The Curvature of the Space of Uncertain Data	50
6.6	Three Geometries	51
6.7	Conclusions	52

7 Aggregation   55
7.1	Why the Least Squares Method (Frequently) Works	56
7.2	Aggregation of Uncertain Data	57
7.3	The Second Axiom	59
7.4	Conclusions	60

8 Thermodynamics of Uncertain Data   61
8.1	Thermodynamic Interpretation of Gnostic Data Characteristics	61
8.2	Maxwell's Demon	63
8.3	Entropy ↔ Information Conversion	64
8.4	Albert Perez's Information	65
8.5	Statistical Interpretation of Gnostic Data Characteristics	67
8.6	Between Mediocristan and Extremistan	70
8.7	Conclusions	72

9 Kernel Estimation — 73
- 9.1 Parzen's Estimating Kernel — 73
- 9.2 Gnostic Kernel — 74
- 9.3 Scale Parameters — 75
- 9.4 Conclusions — 77

10 Probability Distribution Functions — 79
- 10.1 Probabilities — 79
- 10.2 Data Domains — 80
- 10.3 Tasks Solvable by Distribution Functions — 82
- 10.4 The Estimating Local Distribution — 84
- 10.5 Quantifying Distributions — 85
- 10.6 Empirical Distribution Function and the Fit — 86
- 10.7 Some Applications of Distribution Functions — 90
  - 10.7.1 Revealing Historical Information — 90
  - 10.7.2 Hypotheses Testing — 95
  - 10.7.3 A Large Survey of Chemical Pollutants — 96
- 10.8 The Homogeneity Problem — 99
- 10.9 Conclusions — 104

11 Applications of Local Distributions — 107
- 11.1 Enrichment of the *EGDF*-Analysis — 107
- 11.2 Revealing Inner Structure of a Data Sample — 108
- 11.3 Marginal Analysis — 109
- 11.4 Information Capability of Data — 112
- 11.5 Interval Analysis — 113
- 11.6 Diversity of Samples — 115
- 11.7 Conclusions — 117

12 On the Notion of Normality — 119
- 12.1 Normality of Data — 119
  - 12.1.1 Statistical Approach — 120
  - 12.1.2 Empirical Way in Clinical Practice — 121
  - 12.1.3 Similarity-Based Reference Values in Economy — 122
  - 12.1.4 Fuzzy-Set Approach — 123
  - 12.1.5 Automatic Warning and Emergency Systems — 124
- 12.2 Requirements to Ideal Estimation of Bounds of Normality — 124
- 12.3 Elements of Gnostic Solution of the Normality Problem in a One-Dimensional Analysis — 125
- 12.4 Critics on the Identity Gaussian $\equiv$ Normal — 127
  - 12.4.1 Re-definition of Normality — 127
  - 12.4.2 On a Still Daydreamed Research Project BONUS — 129
- 12.5 Conclusions — 132

13 Applications of Global Distribution Functions — 135
13.1 Global Distribution Function — 135
13.2 Comparison of Global with Local Distribution — 139
13.3 Two Didactic Stories — 140
13.4 Conclusions — 142

14 Data Censoring — 143
14.1 Uncensored Data — 143
14.2 Left-Censored Data — 144
14.3 Right-Censored Data — 146
14.4 Interval Data — 149
14.5 On an Unknown Limit of Detection — 150
14.6 Examples of Surviving — 151
14.7 Non-Standard Application of Data Censoring — 153
    14.7.1 Data and Psychology — 153
    14.7.2 Three Aspects of Data Interpretation — 154
14.8 Conclusions — 157

15 Gnostic Thermodynamic Analysis of Data Uncertainty — 159
15.1 Gnostic Data Calibration — 159
    15.1.1 Real Data for Examples — 159
15.2 Data Calibration — 163
    15.2.1 LS-Optimal Numerical Operators — 164
15.3 Calibration of the NIST12 Data — 164
15.4 Calibration of the NIST37 Data — 167
15.5 Conclusions — 169

16 Robust Estimation of a Constant — 171
16.1 Gnostic Data Aggregation Principle Used in Estimation — 171
16.2 Scale Parameter — 172
16.3 More on the Gnostic Data Aggregation — 173
    16.3.1 Example — 174
    16.3.2 Example of Robust Estimation of the Mean of Multiplicative Data — 176
    16.3.3 Robust Estimation of the Mean of Simulated Data — 177
16.4 Conclusions — 179

17 Measuring the Data Uncertainty — 181
17.1 Shortly on the Standard Approach — 181
17.2 The Need of Objective Measuring the Variability — 183
17.3 The Triplication of the Mean Values — 183
17.4 The Need of a Unit of Uncertainty — 184
17.5 The Error of a Mean — 186
17.6 Examples — 186
    17.6.1 Swiss Fertility and Socioeconomic Indicators (1888) Data — 187

		17.6.2 Financial Statement Analysis	188
		17.6.3 Weather Parameters	189
		17.6.4 An Important Medical Parameter	190
		17.6.5 Non-homogeneous Data	191
		17.6.6 Parameters of Uncertainty	194
	17.7	Discussion on Different Means	196
		17.7.1 Re-definition of Variance	196
	17.8	Conclusions	197

18 Homo- or Heteroscedastic Data   199
   18.1 Decision Making 199
   18.2 Examples 200
   18.3 Conclusion 203

19 Gnostic Multidimensional Regression Models   205
   19.1 Formulation of the Robust Regression Problem 207
   19.2 Additive and Multiplicative Regression Models 214
   19.3 Comparison of Robust Regression Models 215
        19.3.1 Statistical Methods for Comparison 215
        19.3.2 Robust Regression in Mathematical Gnostics 216
        19.3.3 Data for Comparison 218
        19.3.4 Criteria for Evaluation of Methods 219
        19.3.5 Results of Comparison 220
        19.3.6 Discussion of the Results 221
   19.4 The Explicit and Implicit Regression Models 222
   19.5 Examples 225
   19.6 Homogeneity of an MD-Model 229
   19.7 An Important Multidimensional Model 230
   19.8 Applications of the Robust Regression Models 233
   19.9 Conclusions 233

20 Data Filtering   235
   20.1 Filtering 235
   20.2 Total Data Variability and Its Components 235
   20.3 Filtering by Regression 236
   20.4 Filtering Effect of Proper Data Aggregation 237
   20.5 Improving the Matrix Quality 238
   20.6 Cleaning of Matrices 240
   20.7 Conclusions 242

21 Decision Making in Mathematical Gnostics   243
   21.1 Datacratic Decision Making in Mathematical Gnostics . . . 244
   21.2 Conclusions 246

22 Comparisons — 247
22.1 Comparisons of Measurement of Toxicity 247
22.2 Comparison of Measurement of Concentration of Cannabinoids 249
  22.2.1 Comparison of Multiplicative Errors 251
22.3 Requirements to the Advanced Comparison 252
22.4 Preparing Data for Analysis 253
22.5 Analysis of Measurement Errors 254
  22.5.1 Characterization of Data 254
  22.5.2 Comparison by Parameters 254
22.6 Conclusions 256

23 Advanced Production Quality Control — 257
23.1 Exploratory Analysis 257
23.2 Automation of the Exploratory Analysis 258
23.3 On the Necessity of Data Inspection 259
23.4 Data Certification 259
23.5 Example of Advanced Quality Control 261
23.6 Homogeneity and Outliers 264
23.7 Estimation of Left-Censored Data 265
23.8 Data Certification and Interval Analysis 267
23.9 Comparison of Laboratories 270
23.10 Conclusions 271

24 Robust Correlation — 273
24.1 Correlation via Distribution Functions 274
24.2 Correlations by Means of Regression 275
24.3 Correlation and Filtering 276
24.4 Autocorrelations 277
24.5 Conclusions 279

25 General Relations — 281
25.1 Relations Considered in Mathematical Gnostics 282
25.2 Robust Curve Fitting 286
25.3 The Experimental Mathematics 286
25.4 Visualization of a Matrix 290
25.5 Critical Points 292
  25.5.1 Relations in Biology 294
  25.5.2 Relations in Technology 296
  25.5.3 Relations in Meteorology 297
  25.5.4 Auto-Relations 302
25.6 Conclusions 305

Bibliography — 307

Index — 315

Preface

*M*otto:
"Mathematical gnostics is a deterministic theory of indeterminism."

Jan Amos Víšek
Charles University, Prague, Czech Republic

Preface I: Strangeness of the Approach

The motto of introducing the book is rather a bon mot than a paradox. It realistically characterizes the approach of mathematical gnostics to data uncertainty. Many consider uncertainty as something indeterministic, i. e. as events not having a precedent cause. Unlike this, data taken in this book are completely determined by measurement and by conditions influencing the measured values. Not all factors determining the observed value are directly measureable but they have a material nature and they are subjected to Laws of Nature. As such, they can be modeled mathematically and estimated by the data analysis starting with the uncertainty of an individual data item. Statisticians working only with "sufficiently" large data samples feel such aims as strange in a way similar to perception of the famous E. A. Abbott's two-dimensional Flatland[5] by three-dimensional people. There is a classical comment to the problem of strangeness cited by E. A. Abbott in his book:

> Horatio: "O day and night, but this is wondrous strange
> and as a stranger give it welcome."
> Hamlet: "There are more things in heaven and earth, Horatio,
> than are dreamt of in your philosophy."[1]

The idea of a theory of uncertainty of individual data (and of uncertainty in small data samples) is strange from the point of view of statistics dealing with mass data. However, information contained in a collection of data is undoubted. Therefore, information contained in a single data item is to be accepted as well. It is then reasonable to look for a mathematical model of the uncertainty of the individual data item and for the uncertainty-information relation like in the "collective" ("macroscopic") statistical case. There is

---
[1] W.Shakespeare, Hamlet, act 1, scene 5.

a significant historical precedent in physics demonstrating the way from a (Newtonian) microscopic model of a single gas molecule's kinetics to macroscopic characteristics of gas kinetics (pressure, temperature, and Maxwell's probability distribution of molecules' velocity). However, which should play the role of "mechanics of individual uncertainty"? One candidate is surely excluded, the statistical Law of Large Numbers, although—as exposed below—there are special conditions (weak data uncertainties), under which gnostic variables and results coincide with the statistical ones. But in physics there is a notion of entropy as a measure of the uncertainty which might be worth of investigation in connection to data uncertainty.

Preface II: Why Gnostic?

A general cybernetic model of recognition of a real object (called *gnostic cycle*) can be represented by a closed cycle consisting of two branches namely

1. observation and
2. action.

This idea firstly appeared in [9]. In the first phase, an observer observes the object to work out a decision on how to change the object and/or its environment to satisfy some conditions. The second phase is a feed-back that implants the control decision.

There was a good reason to call this cycle *gnostic* because this word means *knowledge* in Greek and is widely used in expressions like diagnosis, prognosis a. o. Statistics can be considered as *agnostic* because of its dependence on a priori assumptions on data models rather than on knowledge obtained by careful data analysis. The adjective "Mathematical" should distinguish the mathematical gnostics from the historical religious and philosophical applications of the notion of Gnostics.

Unlike the general idea of the mentioned gnostic cycle, the first (observing) phase is called *quantification*, because it serves to bring to the observer the quantitative information on the state of the observed Nature obtained by counting or measuring. This information is not true because of inevitable disturbances of the process. This is why the second phase called *estimation* follows to estimate the true values of the observed quantity.

The gnostic cycle obtains the name Ideal Gnostic Cycle in mathematical gnostics because it is shown to have some special features when it implements the principles of mathematical gnostics.

Let us take part in an exciting trip into the realm of uncertain events where unexpected rules enable the true quantitative features of observed objects to be uncovered as best as possible from the veil of uncertainty. The theoretical chapters can be left out by readers who do not like mathematical argumentation and they can go directly to numerous applications which could be of interest to everybody. But doing this, they would be made to simply believe that everything is supported by valid, rational consideration.

Preface III: History

The mathematical gnostics cannot be taken as a news. The first publications in print appeared in 1984 and was presented at an international conference of International Federation of Accountants by its author, Pavel Kovanic, who was working as a researcher in the Institute of Information Theory and Automation of the Czechoslovak Academy of Sciences, Prague. The new ideas were met by statisticians not as friendly as is "normal" in science in cases of a new paradigm drastically differing from the accepted "main stream." Nevertheless, a small group of supporters was formed within the Institute that would take part in the formulation of and making precise new ideas. Many thanks are due in this connection to I. Kramosil, J. Šindelář, J. Jarušek, J. Ježek, J. Novovičová, A. Tuzar, and especially to A. Perez, without whom it would be impossible to continue in development of the new approach to uncertainty. A strong support came from many fields of applications, where new algorithms were found to be more suitable for solving problems of praxis than the statistical methods. The applications to economic problems and especially to financial statement analysis in collaboration with D. Hrdinová (later Kovanicová) resulted in a series of publications and repeatedly issuing books on accountancy and financial analysis. The "velvet" revolution enabled P. Kovanic to submit the thesis for the D.Sc. degree. The thesis was sufficient for Z. Wagner for creating his independent version of gnostic software in the Octave language which was many times applied in the analyses of his work and of works of his colleagues in the Institute of Chemical Processes of Czech Academy of Sciences in a large series of articles published in top journals. The availability of internet enabled to contact American professor M. B. Humber, to arouse his interest and to collaborate with him in preparing the book [29]: P. Kovanic got the data from Prof. Humber, analyzed them with his programs, and wrote the first version of each chapter in "Czenglish" for a discussion and for editing it in American English for Prof. Humber. Unfortunately, shortly after finishing the book in the end of 2003, M. B. Humber died. The book [29] was then made available on internet after the agreement with M. B. Humber's family.

Historical development of gnostics and its applications can be traced in a series of publications: [41, 44–47, 54, 55, 84, 86, 99], in scientific reports [42, 43], in doctor (DrSc.) thesis [62], in proceedings of national and international scientific meetings [12, 15, 47, 49, 59–61, 63–66, 70, 71, 75–78, 80–85, 90], and in a series of books [16–18, 67]. Special projects applying the mathematical gnostics are also to be mentioned [68, 69, 72–74]. Development of the gnostic software accompanied the development of the theory and its applications ([13, 48, 50–53, 56–58, 88]) resulting in the first commercially available Gnostic Analyzer ([79] and [87]) usable in Microsoft Windows systems. Later development was using the S-PLUS[2] an then by using the environment of the R-project. The information on the R-project is available here ([96]). A serial application of gnostic methodology for the treatment of environmental and medical data started in 2000 in the Institute of Public Health, Ostrava, and continued by participating in three research projects of the European Union: MAGIC (2005–2008, [36]), 2-FUN (2008–2011, [4]), and FOKS (2009–2012, [3] and [1]).

A summary of ideas, on which the gnostic theory is based, its main statements and its applications were presented in [29].

Conclusions

Unlike mathematical statistics which deals with a "sufficiently" large data samples, mathematical gnostics develops the theory and algorithms of individual data and includes small data samples. The data uncertainty is considered as an effect of the real conditions of measuring and as events subdued to laws of nature. As a new paradigm of data variability, the mathematical gnostics has been developing since the end of 1970s, 20th and applied in a number of research projects.

---

[2]S-PLUS is a registered trademark of the TIBCO Software Inc.

Introduction

The uncertainty of quantifying (counting or measuring) the real quantities has been attracting attention of scientists and technologists for ages. Its significance is increasing in recent development of society which is more and more dependent on information technology. This can be documented by the activity of the conferences of (Information Processing and Management of Uncertainty in Knowledge-Based Systems (IPMU)) where tens of both mathematical and technical approaches to uncertainty are discussed.[3,4] A general conclusion can be drawn out from this activity: there exists a deep dissatisfaction with the statistical methods that were dominating as the anti-uncertainty weapon for hundreds of years. This point can be supported by discussions taking place within the field of statistics: There are altogether seven classes of theories of probability based on different paradigms described in detail and analyzed in [30]. The conclusions drawn therein are far from optimistic:

Clearly much remains to be understood about random phenomena before technology and science can be soundly and rapidly advanced. It is not only the laws of today that may be in error, but also our whole conception of the formation and meaning of laws.

The scope of criticism can be further extended by mentioning the problem of robustness. Hundreds of algorithms produced by robust statistics appeared to work on some special data fail otherwise.

One reason of this state can be found in historical development of statistics. Its origin happened at the ancient time under the need of necessity to implement taxes and to summarize information for ruling states. Probability attracted attention of people in connection with gambling for hundred of years. However, how can be explained that the probability—the kernel of statistics—after its long history not only has seven mathematical models but also this Fine's evaluation:

The many difficulties encountered in attempts to understand and apply present-day theories of probability suggest the need for a new perspective.

---

[3]So, e. g., 127 papers on 34 sessions was presented at the occasion of 16th conference IPMU 2016 in Eindhoven.

[4]There is a traditional inexactness in using the term *uncertainty*. Real data are always modeled as a superposition of two components, the true data and uncertainty. The data variability—as the object of statistical and other studies—is caused not only by impact of uncertainty but also by changes of true data. This is a reason of using the more general term *total data variability* instead of the single *uncertainty* when assuming the possible changes of true values.

Conceivably, probability is not possible. A careful sifting of our intuitive expectations and requirements for a theory of probability might reveal that they are unfulfillable or even logically inconsistent. Perhaps the Gordian knot, whose strands we have been examining, is best cut. However, where would such a drastic step leave the world of practice?

This is a straightforward call for a new paradigm. When looking for a kernel of statistical thinking, one sooner or later comes to the Law of Large Numbers the notion of which can be found already in the works of Italian mathematician Gerolamo Cardano (1501–1576) who stated without proof that the accuracies of empirical statistics tend to improve with the number of trials[31]. However, according to the Czech philosopher Bohuslav Blažek [10], science is developing by the explication of concealed assumptions. It should not mean that neither G. Cardano nor his famous followers J. Bernoulli, S. D. Poisson, P. L. Chebyshev, and others knowingly concealed the fact that the additive aggregation of empirical statistics can be used only under the condition of a weak uncertainty to escape contradictions with the—at that time unknown—relativistic mechanics. They could not even know, that the Euclidean geometry they were using was valid only under the "concealed" assumption of Euclides that speed of light was finite. But this finiteness is recently known and the relativistic mechanics finally accepted but without a response of statistics. This return us to the Blažek's statement, which can be read as a serious warning: When a concealed assumption happens to become an accepted knowledge, science cannot ignore it, because without putting it in use it cannot develop. But this is the case of statistics which missed not only the Einstein's revolution in physics, but also three other scientific revolutions of 19th Century: Helmholtz's measurement theory, Riemannian geometry and the thermodynamic revolution with the Clausius's introduction of entropy. All they came with ideas usable for creating a new—essentially thermodynamic—theory of uncertain data, mathematical gnostics.

But on the other hand, the existence of risk in decision making is beyond all disputes and if the probability is taken as a measure of risk, it simply must exist and the question is what should be the axiomatic cornerstone of its mathematic theory especially if the theory should start with an individual data item and only then extended to data samples. Such a cornerstone has been found in theory of measurement which came with mathematic formulation of the requirements to the every-day activity of people in quantification of things necessary for market, technology, science, health care, and other needs of life.

"Let data speak for themselves" is the idea of an analyst getting data that results from observation, counting or measuring. There exists a lot of factors because of which the data are more or less far from the required true values making them uncertain. The book aims to represent the mathematical gnostics (MG) in double ways:

1. As a tool helping to substitute the old Euclidean-Newtonian way of statistical thinking by the Riemannian-thermodynamic way, and

Introduction

2. As a source of algorithms enabling to get maximum information from the data while relying on the data values only, including individual data items and small samples.

The algorithms are based on the axiomatic mathematical theory closely related to laws of Nature as reflected by such branches of physics like measurement theory, thermodynamics and relativistic mechanics. However, the application of algorithms does not require knowledge of the scientific background of the approach because the descriptions of application methods use only the ordinary language. Moreover, the experience obtained by applications makes the user to be sure on results' trustworthiness. A complete mathematical description of the theory is available in [62] and [29].

Conclusions

The mathematical statistics was the unique weapon in fight against uncertainty of data for the past 100 years. However, the development of information technology revealed the limitations of statistical methods and gave rise to efforts which could improve the statistical methods. Both mass applications of statistics and its own theoretical analyzes have shown that the present paradigm of statistical approach does not give chances for extensions of applicability of statistics and they have been brought to call for a change of paradigm of uncertainty. Many technical innovations along with attempts of theoreticians appeared to answer the needs of practice. The approach called mathematical gnostics is an axiomatic mathematical theory supported by recent state of knowledge of the laws of Nature which generates algorithms robustly solving the problems of praxis.

Author Biography

Pavel Kovanic (Born 1928) 1950-1955: Studied high voltage technology on the Technical University in Sverdlovsk (recently Ekaterinburg, Russia). 1956-1970: Researcher, head of a scientific department at the Nuclear Research Institute of the Czechoslovak Academy of Sciences. 1970-1995: Scientist at the Institute of the Automation and Theory of Information of the Czech Academy of Sciences. 1995-2018: As a retired scientist participated as a scientific consultant on several research projects including grants of European Union.

1
Introductory Kindergarten

The aim of authors as well as of the publisher is to provide a useful book to all types of people. The first condition is the understandability of such a product. The uncertainty is a complex phenomenon of the Nature and to manage it requires mathematical means. Let the reader take it easy, but it cannot be assumed that everybody is ready to consume products of mathematics with a pleasure or at least with understanding. This is why the authors decided to write a short introduction explaining some of the most important elements of mathematics for those needing it. Highly knowledgeable readers are recommended to leave this chapter and to proceed further.

1.1 Elemental Notions

The most elemental object of mathematics is a **number**. It may have an abstract nature, being defined in mathematics. But we shall use its different role, because it can express the amount of things, their quantity. There is therefore a relation between something existing in the real word and the number. We shall call this relation **mapping** and the observed, counted or measured products of this mapping **data**. A **scalar** is a single number, an image of a real quantity in mathematics. A collection of data may have a form of an arbitrary **set** or may be organized in a **row** or in a **column** forming a **vector**. A basic feature of a vector is its **length** equal to number of vector's data. The notion of data may be singular or plural. When speaking on the former case, the notion of **data item** will be used. Several vectors of the same lengths can form a numeric **matrix**. Its basic characteristic is its **dimension** written as the double (r, c) where r is the number of rows and c the number of columns. There exists an important notion of **operation** denoting one of a great number of manipulations with numeric objects. The basic ones are **additive** and **multiplicative** operations that split the set of all numbers into two sub-sets of additive and multiplicative numbers. A set endowed with an operation is called **structure**. The structure of additive numbers is formed by numbers that can be added or subtracted. Zero is an important element of this structure. The structure of multiplicative numbers includes numbers that can be multiplied or divided. Unit is an important element of this structure.

Addition and multiplication is defined for data in the interval $(-\infty, \infty)$, but the notion of **additive data** will be applied to data in the interval $(-\infty, +\infty)$ and the notion of **multiplicative data** will relate with the data in the interval (ϵ, ∞) with $\epsilon > 0$.

1.1.1 Abelian Group

There exist two forms of special structure needed in the sequel: **Abelian group**. An Abelian group is a set, A, of elements of arbitrary nature, together with an (binary) operation $\circ$ that combines any two elements a and b to form another element denoted $a \circ b$. The symbol $\circ$ is a general placeholder for a concretely given operation. To qualify as an Abelian group, the set and operation, $(A, \circ)$, must satisfy five requirements known as the Abelian group axioms:

Closure For all a, b in A, the result of the operation $a \circ b$ is also in A.

Associativity For all a, b and c in A, the equation $(a \circ b) \circ c = a \circ (b \circ c)$ holds.

Identity element There exists an element e in A, such that for all elements a in A, the equation $e \circ a = a \circ e = a$ holds.

Inverse element For each a in A, there exists an element f in A such that $a \circ f = f \circ a = e$, where e is the identity element.

Commutativity For all a, b in A, $a \circ b = b \circ a$.

The set of additive numbers is an Abelian group where the zero is the identity (neutral) element. The set of multiplicative numbers is also an Abelian group with the unit as the identity element.

The elements of an Abelian group may be numbers but their nature may be quite different, e. g. quantities of real objects, their movements like shifts rotations, etc.

Members of a structure can be aggregated to form a class which unifies their features. The aggregation operation can be additive (like forming a flock of sheeps) but not necessarily. There are nonadditive aggregation operation as well.

1.1.2 Variability

The sets of observed real numbers are rarely constant because they can be changed by movements of the object and the impacts of uncertain nature which will be called **uncertainty**. The variability—especially that caused by uncertainty—is measured in statistics by **variance** which is equal to mean square deviation from the average and **standard deviation** (STD)—the root square of variance. Their advantage is in connection with the so called **normally! distributed** processes (processes the distribution of which is Gaussian). Gnostics introduces more general measures of variability which tend to the statistical ones in special cases of weak variability. Two important notions connected with the variability exist, the **homoscedascity** and **heteroscedascity** and are used in statistics for data having constant or variable variance. In gnostics these notions will be used for data modeled with constant or variable scale parameter (which is a measure of variability). Basic statistical model of observed data consists of two elements of variability, namely true constant or variable value and uncertain component. The variability of the true value results from movements of the observed object caused by some driving forces which can be—at least theoretically—controlled unlike the uncertainty. This is why the **total variance** is in gnostics considered as aggregation of the **controlled variance** and of the **uncertainty**.

1.1.3 Morphism and Invariant

A morphism is a structure-preserving map from one mathematical structure to another one of the same things. When we apply exponential function to an element of the additive Abelian group, the result will be an element of the multiplicative Abelian group. The zero will be transformed as unit: the mapping will be structure-preserving because the logarithmic transformation will map the transformed elements back to its previous forms, the additive structure operation will be transformed as the multiplicative one and backward. This structure operation is invariant to exponential/logarithmic transformation. More can be said on this morphism, which is the **isomorphism** or **one-to-one** mapping. This notion is very important because it allows relations between objects to be mapped and important features of the structures to be transferred.

There are two objects that are associated to every morphism, the **source** and the **target**. For many common categories, objects are sets (often with some additional structure) and morphisms are functions from an object to another object. Therefore, the source and the target of a morphism are often called **domain** or **data support** and **codomain** or **range**, respectively. A morphism f with source X and target Y is written $f : X \to Y$. Thus a morphism is represented by an arrow from its source to its target. Morphisms are equipped with a partial binary operation called **aggregation**. The aggregation of two morphisms f and g is defined if and only if the target of f is the source of g and the target of the aggregation is the target of g. There are different morphisms differing by their satisfying different combination of additional conditions, e. g. closeness, commutativity, associativity, distributivity, invertibility.

1.1.4 Vector Space

A vector space is a set V on which two operations $+$ and $*$ are defined, called **vector addition** and **scalar multiplication**. The operation $+$ (vector addition) must satisfy the following conditions:

Closure: If u and v are any vectors in V, then the sum $u + v$ belongs to V.

Commutativity: For all vectors u and v in V, $u + v = v + u$.

Associativity: For all vectors u and v in V, $u + (v + w) = (u + v) + w$.

Additive identity: The set V contains an additive identity element, denoted by 0, such that for any vector v in V, $0 + v = v$ and $v + 0 = v$.

Additive inverses: For each vector v in V, the equations $v + x = 0$ and $x + v = 0$ have a solution x in V, called an **additive inverse** of v, and denoted by $-v$.

The operation $*$ (scalar multiplication) is defined between real numbers (or scalars) and vectors by following formulae: If v is any vector in V of length L, and c is any real number, then $c * v = \sum_{i=1}^{L}(c * v_i)$. For all vectors u, v in V, $u * v = \sum_{i=1}^{L}(u_i * v_i)$. The operation must satisfy the following conditions:

Closure: If v is any vector in V, and c is any real number, then the product $c * v$ belongs to V.

Commutativity: For all vectors u, v in V, $u * v = v * u$.

Distributivity: For all real numbers c and all vectors u, v in V, $c * (u + v) = c * u + c * v$. For all real numbers c, d, and all vectors v in V, $(c + d) * v = c * v + d * v$.

Associativity: For all real numbers c, d, and all vectors v in V, $c * (d * v) = (c * d) * v$.

Unitary law: For all vectors v in V, $1 * v = v$.

A set of values that show an exact position in vector space is called **coordinates**. A scalar product is a binary operation associating a scalar value with a pair of the vectors which have the same length. The scalar products can be used to measure distances between points of the space and angles between vectors making thus the **metric space**. The scalar product is defined differently in different geometries as discussed below.

1.1.5 Matrices

A matrix is a rectangular array composed of r rows and c columns that may have their names (**rownames** and **colnames**). The pair of r and c is called the **dimension** of the matrix. A matrix can be formed of numbers, functions,

Elemental Notions 5

and texts as well as of figures. The element of a matrix M is identified by its content and location as $M_{r,c}$. If there is an operation $f(M)$ applied to a matrix, it is applied to all of its elements ($f(M_{r,c})$. Function f may be any operation applicable to the matrix's elements. A numeric matrix defines a vector space by its rows and columns, each interpreted as vectors belonging to a subspace that is defined by an orthogonal vector. The number of mutually orthogonal sub-spaces of a matrix is called the **rank** of the matrix. The rank is not exceeding the lesser of r and c. The basic operations defined on numerical matrices are especially as follows:

Transposition: The transposed matrix denoted by $t(M)$ is created by the matrix M by exchanging its rows with its columns preventing their orders.

Multiplication by a constant K: Implemented by operation $KM_{i,j}$ for all i, j.

Addition/substraction: Denoted $M_1 \pm M_2$, applied as $M_{1,i,j} \pm M_{2,i,j}$ for all i, j.

Product of matrices: The operator of matrix multiplication is denoted by symbols %∗%. Defined for vectors R1, C, R and C2 (where R = C of length K) and matrices $M_{R1,\,C}$ and $M_{R,\,C2}$ as a $R1 \times C2$ matrix $M_1 M_2 = \sum((M_{i,\,k} M_{k,\,j})_{k\,=\,1}^{k\,=\,K})$ for all i, j. This product **is not commutative**.

Eigen decomposition: Defined for symmetrical matrices ($M_{i,\,j} = M_{j,\,i}$ for all i and j) as $M = M_0 D_e t(M_0)$ where $t(M_0) M_0 = E_R$. The columns of M_0 are **eigen vectors**, the D_e is a diagonal matrix of R positive **eigen values** and E_R is the **unitary** matrix of R units. The number R is the matrix's **rank**.

Singular value decomposition: A rectangular matrix M with r rows and c columns (where $r \geq c$) is presented as a matrix product $M_{L,r,c} D_c t(M_{R,c,c})$ where $t(M_{L,r,c}) M_{L,r,c} = E_c$, D_c is a diagonal matrix of c nonnegative **singular numbers** and $t(M_{R,c,c}) M_{R,c,c} = E_c$. The columns of $M_{L,r,c}$ and rows of $t(M_{L,r,c})$ are **left and right singular vectors**. The number of positive singular numbers is the **rank** of the matrix.

Pseudo-inversion: The matrix $M_{c,r}^+ = t(M_{R,c,r} D_c^+ M_{L,c,c})$ where positive numbers on diagonal of the matrix D_c^+ are reciprocal values of the positive diagonal numbers of the matrix D_c and is the pseudo-inverse of the matrix $M_{r,c}$ for which $R \leq c$ is the rank. In a general case the relations $t(M) M^+ M = M$ and $M^+ M M^+ = M_+$ hold. When the rank equals the number of rows r, the special case called **inversion** of the matrix M takes place. For an inversion matrix denoted M^{-1}, equations $M M^{-1} = M^{-1} M = I$ where I is the unit matrix.

As shown in [94] the pseudo-inverse can be used to compute least-squares-minimal numeric linear operators to obtain results of many different linear operations on data series.

1.1.6 Probability Distribution

The probability distribution function is in statistics defined as a function that describes the likelihood of obtaining the possible values that a random variable can assume. Instead of academic notion of random function, mathematical gnostics operates on real variables which may be disturbed by an uncertainty. The values of the observed variable are taken as **quantiles** and the probability of having a value of a quantile is to be estimated. The probability distribution function of a set of quantiles (also called probability domain, or data support), is a set of probabilities of quantiles. The probability density function is a rate of change of probability function which depends on quantiles. The probability distribution function has the cumulative feature and it is not decreasing in dependence on quantiles. It gives to each quantile the probability of not exceeding the quantile's value. Unlike this, the probability density function says what probability is attached to the interval of quantiles (q_2, q_1). Unlike statistics preferring probability function of an a priori assumed form, the gnostic probability distribution is always estimated by using the observed data.

1.2 Sources of Inspiration for Mathematical Gnostics

Dissatisfaction with statistical methods were accumulated during the years of application of statistics to solving the problems appeared in the research activity of authors. More attention was attracted to permanent discussions between different groups of statisticians. The repeatedly cited Fine's book [30] came after a careful analysis of several approaches to statistics reached a conclusion that it led to improvement of the state of statistical affairs: *Perhaps the Gordian knot, whose strands we have been examining, is best cut. However, where would such a drastic step leave the world of practice?* A chance for such a cut appeared by the first publications on mathematical gnostics in 1984 ([44],[45] and [46]) which resulted in series of successful applications, but with a zero attention of the statistical mainstream. Mathematical gnostics build its axiomatic models of uncertainty differently than statistics using approaches and results of several scientific branches:

1. Theory of general systems.
2. Theory of measurements.
3. Geometries including some non-Euclidean ones.
4. Relativistic mechanics.
5. Thermodynamics.

All these sources started their journey in 18th century and did not find a proper reflection in the unique science of uncertain events, statistics. It is not necessary to be an expert of the indicated science branches. It is sufficient to accept their main statements to understand the gnostic approach to uncertainty.

1.2.1 Theory of General Systems

The idea of a gnostic system for recognition, consisting of two phases, namely observation and a corrective feedback that has been published in a book on theory of cybernetic system [9]. The purpose of the feed-back was special: to control an object to compensate the observed deviation from the state required by the observer. In a more general setting, the uncertainty was introduced into the observation phase, and instead of the corrective feedback the estimation of the unknown true quantity of the observed object was considered. The name **gnostic** for the system was taken over because of its Greek origin (knowledge) applied in frequently used notions like diagnostics, prognostics, agnostic, and others. To distinguish the word from its connection with a religion the adjective **mathematical** was added.

1.2.2 Theory of Measurement

The development of the measurement theory was started by H. von Helmholtz [105] who applied mathematics to characterize the quantification of real objects as a mapping of mathematical structures of real quantities and things into the realm of mathematics. He also had shown the mathematical nature of operation of measurements as a series of mathematical conditions necessary for the consistency of the measurements. Conditions of topology were formulated as series of binary relations like coincidence and foregoing (precedence) being symmetric, transitive and reflexive or non-reflexive. Theory of measurement represents today a scientific basis for the important praxis of measuring [27], [19]. All the conditions of theory of measurements were unified in mathematical gnostics as a simple axiom of data which was consistent image of structure of real quantities. This axiom was then used to derive the remarkable model of data uncertainty.

1.2.3 Geometries

Geometry is a branch of mathematics dealing with distances, angles, shapes of figures and relations between abstract objects. When a notion of distance is heard, the classical formula $\sqrt{(x^2 + y^2)}$ automatically appears before our eyes as our education concentrated only on the Euclidean geometry without even mentioning the existence of other geometries. It is a square root of the scalar product of the vector (x, y) with itself, its length. It can be written in the form of $\sqrt{(x^2 - (C \cdot y)^2)}$ with an undeterminate quantity equaling in this case $C = \sqrt{(-1)^2}$. The vector obtains the form $x - C \cdot y$ and is then called *pair* or *double* number. It has been proved in [35] that there are just three kinds of the undeterminate C: J, I, and 0, for which $J^2 = 1$, $I^2 = 1$ and $0^2 = 0$ and that all three have their background in physics. Really, if the coordinate x represents space shift and y the time change, then the vector $x - I \cdot y$ serves as a space-time model of an event of classical mechanics, a movement with a velocity much less than time speed. Unlike this, the vector $x - J \cdot y$ is a model of relativistic movement with a velocity close to the time speed. The vector $x - 0 \cdot y$ models a shift in Euclidean space, when time speed is infinite and time coordinate does not play a role at all. The choice between values of J and I decides the choice between geometries Minkowskian and Euclidean. Both are used in mathematical gnostics like versions "Q" and "E" of processes, functions and formulae, but because of the non-linear nature of uncertain objects they are applied in differential form as in Riemannian geometry.

1.2.4 Maxwell's Contributions

Maxwell's equations of electro-magnetism provide worthwhile lecture on the application to both scalar and vector fields to characterize natural processes including the virtual movement of an uncertain value. Maxwell's demon is a virtual modeling of the conversion of entropy to information and back.

1.2.5 Relativistic Physics

As results from the first axiom of the gnostic theory, there exists a close correspondence of the changes in data variability with movements of unloaded relativistic particles. It is sufficient for understanding the gnostic model of variability to accept the facts, that the energy of the particles is modeled by hyperbolic cosines and their moments by hyperbolic sines and for them the relativistic Conservation Law of Energy and Momentum holds.

1.2.6 Thermodynamics

The mystic notion of entropy has been misused in statistics by introduction of different formal definitions not connected to its primary thermodynamic purpose followed by Rudolf Clausius. In 1855 he defined the entropy as integral of ratio of heat changes divided by the absolute temperature. Mathematical gnostics has shown that this definition can be applicable to data uncertainty and also can be favorably used for revealing its fundamental features and the analysis of uncertain data.

1.2.7 Matrix Algebra

For the manipulation of structures of numeric data elements and operations of matrix, algebrae are used assuming that they are known by readers, otherwise they are available in the documentation of r-project [96, 97]. In cases of necessity, the R-language is used for explanations.

1.3 Conclusions

The mathematical gnostics was inspired by

1. historical successes and failures of statistics,
2. critiques inside of statistics,
3. significant scientific contributions of the following in the 19th and 20th centuries:

   - H. von Helmholtz's theory of measurement,
   - R. Clausius's contribution to the thermodynamics,
   - J. C. Maxwell's theory of electromagnetism and its virtual demon violating the second thermodynamic law,
   - H. Minkowski's and G. F. B. Riemann's contributions to geometry,
   - A. Einstein's relativistic mechanics.

None of these sources found an adequate reflection in the statistical mainstream.

2
Axioms

Let $\mathcal{A}_I$, $\mathcal{A}$ and $\mathcal{N}$ be non-empty sets of real numbers, elements of the set R^1. The elements of $\mathcal{A}_I$ will be called *true values* of a real quantity, elements of $\mathcal{A}$ will be the *observed data* and $\mathcal{N}$ the *uncertainties*. The following mappings will be considered:

1. $\upsilon : \mathcal{A}_I \to R^1$
2. $\vartheta : \mathcal{A} \to R^1$
3. $\nu : \mathcal{N} \to R^1$
4. $\sigma : \mathcal{N} \times \mathcal{N} \to \mathcal{N}$
5. $\pi : \mathcal{A}_I \times \mathcal{N} \to \mathcal{A}$

Mappings υ and ν will be called *ideal quantification* and π *real quantification*. They will serve as mappings of objectively existent structures of natural quantities into mathematical structures.

2.1 Axioms of the Data Model

A1.1 Mappings υ, ϑ and ν are one-to-one. A1.2 Mapping ϑ is an isomorphism between the structure $[\mathcal{N}, \sigma]$ and the additive group $[R^1, +]$. A1.3 There exists such a positive number S, that for all $a_0 \in \mathcal{A}_I$ and $n \in \mathcal{N}$, the equation

$$\vartheta(\pi(a_0, n)) = \upsilon(a_0) + S\nu(n) \tag{2.1}$$

holds. The a_0 is the true value of a real quantity and n is an uncertainty. This equation may be rewritten in the numeric form of

$$A = A_0 + S\Phi \tag{2.2}$$

where $A = \vartheta(\pi(a_0, n))$, $A_0 = \upsilon(a_0)$ and $\Phi = \nu(n)$. Equation (2.2) is the model of additive data. Structure of additive models of the type (2.2) is Abel's commutative group. Parameter S is the scale parameter.

Using the exponential transformation

$$Z = \exp(A) \tag{2.3}$$

DOI: 10.1201/9780429441196-2

one obtains
$$Z = Z_0 \exp(S\Phi) \qquad (2.4)$$
where
$$Z_0 = \exp(A_0) \ . \qquad (2.5)$$
The Equation (2.4) is the *Multiplicative Data Model*.

This text is a citation of the doctor (DrSc.) thesis [62] of the first author. These axioms, from which the all statements of the mathematical gnostics are derived, were motivated by the measurement theory, which became a science thanks to H. von Helmholtz, who introduced the mathematical formulations into the requirements ensuring the measurability of real quantities. The more recent state of the measurement theory has been published e.g. as [27]. The significant difference between the measurement theory and [62] is in the data variability and uncertainty—not considered by measurement theory and left to statistics. Data variability is the main object of study in gnostics.

2.2 Applications of Axiom 1

The variable Z_0 introduced in the data model (2.4) deserves a comment. It might be the true value to be observed, but we do not know its value. Instead, we can apply an estimate to it, using the median of the sample, the estimate of the sample's mean, a chosen value of the sample's data item or another estimate of the true value. Equation (2.4) can be rewritten as
$$R = \exp(S\Phi) \qquad (2.6)$$
where
$$R = Z/Z_0 \qquad (2.7)$$
is the multiplicative error of the estimate. We recognize both of its components, because Z is the observed value and Z_0 is our supposedly done estimate. The R thus characterizes the relations between the observed quantity and our estimate of the true value, and the right hand side of (2.6) represents the relative impact of variability on the estimate. The observed value may be constant or variable. Their are different uncertain disturbances on the observation process. The estimate will thus reflect two kinds of variability, the variability of the true value and uncertainty.

Both forms of the quantity R are useful: the form of (2.6) will be shown to be applied for evaluation of the amount of uncertainty, to compute important data characteristics and to answer the question, if the data sample is homoscedastic or heteroscedastic. On the other hand, the R interpreted according to (2.7) enables the making use of the hyperbolic or trigonometric

Applications of Axiom 1

functions in analysis by introducing four data characteristics usable as coordinates of a couple of bi-dimensional spaces called *Q-space* and *E-space* alias *quantifying space* and *estimating space*.

An auxiliary variable R was introduced which has two interpretations: the angular (2.4) and that of multiplicative estimation error (2.5). Using the identities

$$R \equiv (R + 1/R)/2 + (R - 1/R)/2 \equiv \cosh(S\Phi) + \sinh(S\Phi) \qquad (2.8)$$

one introduces the definition of two variables

- f_J ... *quantifying data variability*
- h_J ... *quantifying irrelevance*

where

$$f_J = (R + 1/R)/2 = \cosh(S\Phi) \qquad (2.9)$$

and

$$h_J = (R - 1/R)/2 = \sinh(S\Phi) \ . \qquad (2.10)$$

Accepting the functions f_J and h_J as coordinates of the data space, we also accepted the space geometry: Indeed, Equation (2.9) prescribes the way of evaluation of the multiplicative error R, which originates by the substitution of the observed data value Z instead of its true value $Z0$. Relation

$$f_j^2 - h_j^2 = 1 \qquad (2.11)$$

says that when the angle $S\Phi$ would be a constant and the space would be Minkowskian plane. However, the error R is ordinarily a variable. The metric is therefore Riemannian and—because of variability of the $S\Phi$—the space is curved. However, we shall need an alternative—Euclidean—coordinate system of the data space. Such a system is already on hand, as documented by the equation

$$(1/f_J)^2 + (h_J/f_J)^2 = 1 \qquad (2.12)$$

which defines the required Euclidean trigonometric functions closely connected with the quantifying coordinates:

$$f_I = 1/f_J \ , \qquad (2.13)$$

$$h_I = h_J/f_J \ . \qquad (2.14)$$

This enables two estimation characteristics to be defined:

- f_I ... *estimating data variability*
- h_I ... *estimating relevance*

where

$$f_I = 2/(R + 1/R) = \cos(S\phi) , \qquad (2.15)$$

$$h_I = (R - 1/R)/(R + 1/R) = \sin(S\phi) . \qquad (2.16)$$

Variables f_J, h_J, f_I, h_I will be called *gnostic characteristics*. Equations (2.13) and (2.14) define the isomorphism between quantifying and estimating data characteristics. The couples of hyperbolic and trigonometric functions used as coordinates of uncertain data planes enable further—not only geometric—features to be investigated.

2.3 Data Aggregation as the Second Gnostic Axiom

Statistics uses additive aggregation law applied to data values. This is not a trivial problem as shown in the chapter Aggregation. After a careful consideration of factors resulting from the gnostic data model and, using inspiration of aggregation in physics described there, it is possible to formulate Axiom 2 of gnostic theory: Additive aggregation law for uncertain data is applied to four gnostic characteristics f_J, h_J, f_I and h_I. Validity of this law for quantification characteristics f_J and h_J is discussed in chapter Aggregation. Validity for the estimation characteristics f_I and h_I is supported by the isomorphism between both coordinate systems.

2.4 Conclusions

Inspired by the theory of measurements, several requirements of algebraic nature are required for data to be accepted as proper mathematical images of natural objectively existent structures of quantities: the data should be elements of Abel's commutative group. The model of observed data includes an image of uncertain data component which is unknown and is to be estimated. After the quantification phase of data changing which transforms the image of quantity into mathematics, the estimating phase must follow to estimate the true quantity's value. Both these phases are two-dimensional including images of the true value and of the uncertainty. To trace the data changes in both spaces (quantifying and estimating) four data characteristics as coordinates of the data of virtual movement are defined: quantifying and estimating the variabilities and irrelevances as the errors caused by variabilities. The second gnostic axiom requires the additive aggregation law for all four data characteristics.

3
Introduction to Non-Standard Thought

3.1 Paradigm

A theory is an abstraction conceived to explain or predict reality. It needs to include sufficient information, but also to suppress irrelevant facts. Therefore a model is proposed as a simplified reproduction of significant features of observed reality. Theories evolve over time and are improved or negated as empirical investigation either supports or refutes, what had previously been postulated.

The notion of a *paradigm* is close to that of a *model*, but Thomas Kuhn [98] introduced a special interpretation, which pertains to scientific revolutions. He suggests, that a paradigm represents a collection of generally accepted views, which dominate the thinking of "experts" in a scientific field at some point in the development of a theory. As shown later by Joel Barker [6], the problem of the paradigm is much more universal in its nature and it is one of the most important questions in the development of our every-day life. A paradigm consists of two major parts. It:

1. delimits the boundaries of a class of problems and
2. includes a collection of rules to solve the problems, which exists within the given boundaries.

The acceptance of an existing paradigm results in several advantages: The paradigm

- helps to distinguish between the important and the insignificant,
- offers advice and recommendations as to how to move successfully within the given boundaries,
- aids in communication between its adherents, because they are all familiar with it and use the same notions, terms and language,
- helps in the understanding of changes within its "valid" framework, because it is understood as being "legal" and it does not give rise to suspicions of "heresy" within the domain and it does not lead to conflicts with the "pontiffs," who dominate the field,

- assists in legitimizing activities within its boundaries, thus increasing the number of its adherents and sponsors.

There are also negative features. A paradigm is a "filter," which selects and adapts incoming information to support itself and to eliminate inconsistencies with any new facts. Murphology has two observations on this issue [7]:

1. **Maier's law:** "If facts do not correspond to your theory, get rid of them as fast as possible."
2. **Finagl's credo:** "Science is always right. Do not be confused by facts."

These theses are (sadly) more than a joke; if these conclusions were not frequently real, the acceptance of a new paradigm would be much easier and faster. Blind and uncritical adherence to an existing paradigm often results in narrow-mindedness and to an erroneous conviction, that everything successful in the past must be successful in the future, because the future is nothing more than a simple extrapolation of the past. Paradigm also plays a harmful role in specialization of scientific journals protected by an impenetrable wall of reviewers ready to reject each article bringing a new idea critical to "blessed" points of view.

In defense of maintaining the old order, a change in the paradigm might encompass large risks:

- At the moment, when a revolutionary paradigm is accepted, a great deal of the built-up intellectual or spiritual "capital" of those who supported, nurtured and maintained the supplanted paradigm is lost, and nearly everyone starts from zero once again.

- New paradigms ordinarily appear at the boundaries of several scientific fields, which are not familiar to the "priests" of the old paradigm. Younger scholars with fresh new knowledge, and newcomers from the new "neighboring" fields are favored.

- As a potential revolution of the paradigm develops, it is not sure, who will win. Many will prefer to wait on the sidelines to see, which way the wind blows, before putting their necks on the chopping block or opening themselves to criticism.

- The old paradigm is rarely completely refuted; this permits the established ideas to continue to be harvested until the new ones are completely established. Moreover, some new paradigms are more general than the old ones and include the former as valid special cases. This, of course, does not apply to conflicts between hostile paradigms such as between social systems or between paradigms, which exclude each other (such as the Ptolemaic versus the Galilean paradigms).

- Occasionally, a seemingly new and better paradigm appears to be more a fashion or a fad than a well justified innovation (e.g. many slimming cures, following the "herding instinct" in jogging, etc.).

- "New" is not automatically the same as "progressive" or "better." History is replete with many new ideas or discoveries, which have lead to dead end roads or U-turns, e. g. *DDT*, cheap and safe nuclear energy, etc.
- Some paradigms, especially those related to the use of power may be highly dangerous, (e.g. Hitler's "Blitzkrieg" or religious and political fundamentalism).

Having considered the advantages of continuing to subscribe to an old paradigm with the risks of accepting new or revised ideas, one can develop a better understanding of conservatism in thought and general resistance to change.

It is logical to ask, why such a philosophical and psychological problems are dealt within a book, which aims to contribute to data analysis. The answer is, that to analyze, one needs data and analytical methods. Real data may contain strong uncertainty. Hence, the methodology, which will be applied, must be able to cope with the inherent uncertainties. There are different paradigms of uncertainty; to select the most suitable, it is necessary to also consider, how the uncertainty should be measured, i. e. which geometrical paradigm is to be applied.

The mention of a geometrical paradigm is not random here. Data uncertainty is reflected in data errors. These are distances of the observed data value from a true value. Distances are non-trivial objects of geometry and different geometries measure distances differently.

3.2 Statistical Paradigms

When using the word *statistics*, one must distinguish between two substantially different meanings:

1. numbers, that have been collected in order to provide information about something and
2. the science of collecting and analyzing these numbers.

The numbers cited in both definitions are *data*, i. e. outputs of the quantification process, which maps real quantities. A quantitative depiction of reality would be impossible without data. The statistical activity described by the former definition is an absolutely necessary part of all methods used to obtain quantitative information. The latter meaning also defines an objective of mathematical statistics, but it does not imply its uniqueness as a tool for data analysis.

The use of the plural in the heading above might be a shocking revelation for one, whose acquaintance with statistics is via the most popular paradigm, that of *relative-frequency* statistics. It is based on one of the oldest paradigms, which is closely connected with games of chance such as dice, cards or roulette. The *relative frequency* (number of successes divided by the number of trials) characterizes the success of repeating a given (*random*) experiment under fixed conditions. The thrust of this paradigm is described by [30] as follows:

1. In many important cases relative frequencies appear to converge or stabilize, when the random experiment is repeated a sufficient number of times.
2. This apparent convergence is an empirical fact and a striking instance of order in chaos.
3. The apparent convergence imputes a hypothesis, that the relative frequency of outcomes in as yet unperformed trials of an experiment can be extrapolated from the observed relative frequency of trials already run.
4. Probability can be interpreted through the limit of relative frequency and assessed from relative frequency data.

This statistical paradigm is not unique. There are at least seven classes of theories of probability based on different paradigms, which are described in detail and analyzed in [30]. The pessimistic conclusions drawn therein were already cited above.

Perhaps this sad state of affairs can be interpreted as a call for a good non-statistical paradigm to assess uncertainty.

3.3 Statistical Data Weighing

A statistical theorem is useful to be cited to demonstrate the statistical way of thinking. It relates to chapter "Direct observation of differently precise events" of the old book [101]. Theorem 5.2.1. based on the maximum likelihood method states following: Given n independent and normally distributed observations x_i with distributions

$$x_i \in N\left(a, \frac{\varsigma}{\sqrt{(p_i)}}\right) \tag{3.1}$$

where quantities a and ς are unknown and p_i are reciprocally proportional to the variances $D(x_i) = \frac{\varsigma^2}{p_i}$. Then the estimate

$$\hat{a} = \frac{\sum_i^n p_i x_i}{\sum_i^n p_i} \tag{3.2}$$

is an unbiased and efficient estimate of the mean a with the variance

$$D\left(\hat{a} = \frac{\varsigma^2}{\sum_i^n p_i}\right) . \tag{3.3}$$

Non-Statistical Paradigms of Uncertainty

It also results from the theorem that a collection of normally distributed events with different variances is distributed normally as well. We are not going to criticize the expectation of such a collection on practice neither the difficulties of a decision which individual observation belongs to each subsample but the basic idea can be doubted: "you belong to a family/class/race whose worth/fault (variance) is known, your personal worth/fault is the same and your individual weight is determined by the weight of your collective and not but your individual worth/fault." The opposite way of thinking is that mathematical gnostics considers the individual features of each individual data item.

There also is a natural idea followed by this theorem: the weight of data must be the less the strongest the data uncertainty is. However, the uncertainty must be individual of each data item and not collective like the variance of the whole sub-sample.

Another statistical implementation of the data weighing is the Iterated Weighted Least Squares Method based on influence functions [95]: A robust multidimensional regression model can be obtained by the application of influence functions to weigh each equation of an equation system while iteratively looking for the solution of the system. Many versions of influence functions are available in the literature robustly working in application to some data while failing on others. Nether in this case are the data weighed individually. The weight given to an equation is determined by the errors of the whole equation.

3.4 Non-Statistical Paradigms of Uncertainty

Many problems of a theoretical nature have given rise to new attempts to reconsider various statistical paradigms. The rapid development of computers after World War II enabled existing statistical methods to be applied to real problems to an extent never thought possible before. However, results have been far from satisfactory. This may be explained by the fact, that those statistical methods are products of mathematics; and as such, if they were developed from non-conflicting assumptions in a consistent manner, they cannot be wrong. If a mathematical statistics methodology fails, when it is applied to real data, it is necessary to look for the cause in the conflict between the theoretical assumptions and the real nature of the data. A statistician may warrant his methodology to one, who requests an analysis, only if he in turn provides the statistician with a warranted statistical model of the data. Statisticians ordinarily make the requester responsible for the choice of data model. *The stumbling-block is, that one very rarely knows the statistical model of real data.*

Both theoretical and practical problems with statistical paradigms have led to a fast development of methods based on alternative, non-statistical paradigms. As outlined in [2] several of these methods are being applied in forecasting and decision making in the financial markets. The use of methods based on pattern recognition, neural networks, fractal geometry, deterministic chaos, fuzzy logic, genetic algorithms and non-linear dynamic theory are discussed. It seems, that there is an informal but ferocious "race" running for alternative paradigms of uncertainty. The broad spectrum of non-statistical methods is dealt with every two years by conferences (International Processing and Management of Uncertainty (IPMU)). (The Spanish city Cádiz was selected for the 17-th conference IPMU 2018). It is not the aim of this text to deal with the multitude of non-statistical paradigms of uncertainty. Instead, it shall concentrate on a single participant in the "race," on the *gnostic* paradigm, which was absent from the roll call given in [2].

3.5 On the Need of an Alternative to Statistics

The historical achievements of statistics, especially in physics, has justified the consideration of this methodology for use in the analysis of many phenomena in different fields of science and practice. So, e.g. theories of statistical thermodynamics, chain fission reaction and neutron slow-down and diffusion yielding precise engineering calculations for nuclear reactors constituted some of the unchallenged successes of the statistical approach. However, is this a sufficient reason to expect, that the application of the same principles will yield equally successful results, when applied e.g. to economics? Because economic processes are substantially different from physical ones, it is not likely. Economics produces all material means for the mankind. Let us consider this important field of practice in more detail. Benjamin Graham, the father of "fundamental" investment analysis stated [11]:

> *...The art of investment has one characteristic that is not generally appreciated. A creditable, if un-specular, result can be achieved by the lay investor with a minimum of effort and capability; but to improve this easily attainable standard requires much application and more than a trace of wisdom. If you merely try to bring* just *a little (emphasis added) extra knowledge and cleverness to bear upon your investment program, instead of realizing a little better than normal results, you may well find that you have done worse.*
> *Since anyone—by just buying and holding a representative list—can equal the performance of the market averages, it would seem a comparatively simple matter to "beat the averages"; but as a matter of fact, the proportion of smart people who try this*

and fail is surprisingly large. Even the majority of investment funds, with all their experienced personnel, have not performed so well over the years as has the general market ... there is strong evidence that their calculated forecasts have been somewhat less reliable than the simple tossing of a coin.

Although there is no reference to specific forecasting methods, it can be inferred, that the word "calculated" refers to the mathematical methodology of statistics, that has almost exclusively dominated econometrics for decades. Among other pertinent critiques of the statistical approach to economic problems, following remarks by Los [8] deserve to be mentioned:

... It is clear to most people, that economic forecasting still amounts to little more than educated guessing, despite the aura of precision created by computerized models of the economy.
... Scientific economic analysis, in the true sense of these words, still does not exist.
... Since objective modeling has not been practiced, economics as a science has not progressed.
... Recently, simple cost-benefit analysis has created strong financial incentives to obtain better and more accurate economic forecasts in the private sector. But, paradoxically, the main obstacle to this progress in economics is the conventional pseudo-scientific methodology of econometrics adopted in the 1940's and 1950's. The conclusion is clear: first the problem of objective identification from noisy data has to be solved.

Professor R. E. Kalman, who made a substantial contribution to cybernetics with his famous filters, expressed his view of the issue as follows [21]:

Statistics is not science, but a kind of pre-science, a pseudo-science, a "gedankenscience."[1] Perhaps it's best called an "ersatzscience."[2]
... Uncertainty in nature cannot be modeled (and therefore must not be modeled) by conventional, Kolmogorov[3] probability schemes, because no such scheme may be identified from real data.
... The trouble is, that probabilities are not identifiable.

---

[1] *Der Gedanke...* the thought (in German). Appears frequently in natural sciences in the word *der Gedanken-experiment*—a thought experiment not really performed, but obeying an a priori given system of laws. It is in use without translation in many languages. The most popular application of such an approach was the A. Einstein's cosmic elevator used in the General Theory of Relativity.

[2] German word *der Ersatz* means a not quite perfect substitute, an artificial Christmas tree or a "hamburger" made from soy beans.

[3] A. N. Kolmogorov—Russian mathematician (1903–1997), who developed in the 1930's the most commonly accepted version of probability theory.

Even the name of the scientific meeting, where Kalman's contribution was presented, could be interpreted as symptomatic—*Foundation Crisis in Econometrics within the Standard Statistical Paradigm*. As pointed out in [8], the criticism of current methodologies of data treatment has long ago left academia for the popular press, for instance in the Wall Street Journal [20]:

Fickle Forecasters. How Three Forecasters, After Crash, Revised Economic Predictions.

and [14]:

Into the Void: What Becomes of Data Sent Back From Space? Not a Lot as a Rule.

It is not reasonable to reject statistics, because it is a "gedankenscience." The power of mathematics results from that fact, that it is a "gedankenscience," due to its independence from the facts of real life. However, the practical applicability of mathematical or statistical models goes outside the boundaries of a "gedankenscience." Many processes studied in physics are modeled by "gedanken-experiments," because useful models of their behavior are simple enough to be formulated by humans. We can come very close to describing the orbit of the earth relative to the sun using only Newton's gravitational principle and the masses and distances of the earth, sun, and moon. For most purposes, we can ignore the effects of other planets, other stars and air disturbances due to (say) the flight of butterflies. However, in economics, it is not simple to distinguish the perturbations of the data resulting from influences, that (if we knew, what they were) could be ignored, and the essential ones. It is impossible to discriminate from the "flapping wings of a butterfly" and "the mass of the sun." Moreover, we have not yet identified anything that remotely corresponds to Newton's laws in economics. Such principles, invariant for all time, may not even exist. Nothing is stationary and replicable in economics. One of the major issues is the independence of events; the collision of two gas particles at a specific point can be considered completely independent of a collision of particles at a distant point. Economic events not only are influenced by economic transactions, but also by seemingly unrelated activities across the globe, which may even cause a strong synchronous reaction throughout the world.

The impropriety of statistical applications to many economic propositions is reflected by the manner, in which many problems are stated. They begin with the assumption: *Let $x_1, \ldots, x_N$ be the N-tuple of i.i.d. random variables.* The idea of independence, as noted above, is probably unsuitable for economic events. *Identical distribution* refers to stationarity and repeatability, which is also a doubtful characteristic of economic data. However, the most discordant is the notion of *randomness*. This is pure agnosticism, a complete abdication of the notion, that the human mind has the ability to discern, confirm, and establish the cause of events.

Returning to economics: are the fluctuations of prices on the stock market random? Ask market experts this in a more specific way: "Was yesterday's change in company X's share price random?" The explanation, (or several explanations) received would suggest, that what occurred, was a necessary consequence of having new public information about X's earnings or prospects, or a change in the discount rate by the Fed., etc. The change might seem random for those, who perceive the market only as a big roulette wheel. Often, no reason can be elicited, and the response could be: "I have no idea;" (read, "I have no information,") rather than, "It was random."

These problems play a role in many other application fields. Can be an epidemic considered as a series of stationary, random and independent events? Can be the quality assessment of a serial industrial product based on such assumptions? However, the main tool to prove statistical statements, the Central Limit Theorem, assumes not only a "sufficient" amount of events, but also randomness, independence, a finite expected value and a finite variance.

A long-time experience leads N. N. Taleb to write a popular book [37], which became the New York Times Bestseller. It distinguishes two kinds of "worlds," namely Mediocristan (where the Gaussian bell curve dominates) and Extremistan (the world of highly improbable events). The aim of this book to show that a third large realm exists like Ordinarystan between them where we live and work.

3.6 Principles of Advanced Data Analysis

The foregoing has prepared the way to introduce the concept of *Advanced Data Analysis*, which will be interpreted as an analysis, which respects the objectivity of data and aims to draw out the maximum information, while letting the data decide the fundamental problems of their treatment. The basic rules of advanced data analysis include following:

1. Do not violate the data by
   (a) subjecting them to unjustified a priori models or distribution functions,
   (b) trimming the data sample, because even outliers may bring an information,
   (c) imposing behavior on them in accordance to non-smooth functions only in thoroughly justified cases,
   (d) ignoring that some of them are outliers causing sample's non-homogeneity,
   (e) not respecting the proper way of data aggregation based on the second axiom of gnostic theory,

(f) quantifying their values, differences and weights by an improper geometry,
    (g) not respecting their finiteness.
2. Make use of all available data by
    (a) including censored data,
    (b) including suspected outliers and giving them their justified weights,
    (c) including suspected inliers ("noise") and giving them their justified weights,
    (d) excluding data only after proving their negligible impact or their invalid origin,
    (e) not overlooking side effects caused by the investigated processes.
3. Let data decide
    (a) their membership in additive or multiplicative commutative group,
    (b) on their homo- or heteroscedasticity,
    (c) the finite bounds of data supports,
    (d) on the outlier/inlier ("membership") problem,
    (e) on sample's homogeneity,
    (f) on structure of non-homogeneous data samples,
    (g) the metric of their space and of its curvature,
    (h) on their own individual weights,
    (i) on their uncertainty by evaluating it using their theory,
    (j) their interdependence within the sample and with other samples,
    (k) on the bounds of their domain,
    (l) on their probability distribution and density functions,
    (m) on the separation of the uncertainty from the variability of true data values in the total variability.
4. Determine the weight of each individual data item by its own value and not only by the weight of the sample to which it belongs.
5. Use statistical methods iff the assumptions, on which they are based, are justified.
6. Do not shun the use of good non-statistical methods, when statistics fails or when its application is not appropriate.
7. Use distribution functions where possible instead of point estimates for data characteristics.
8. Take as similar/comparable only objects, which behave in accordance with the same model.

The Gnostic Concept 25

9. Do not blame randomness for effects. Try to explain the causes of uncertainty by using the data and other available information to minimize the uncertainty.
10. Prefer robust estimation and identification methods over non-robust ones.
11. Select the kind of method's robustness (inner/outer) with respect to any given task.
12. Apply realistic criteria (information/entropy) to optimization.
13. Respect theoretically proved optimal paths for data transformation and estimation.
14. Do not allow oneself
    - conservatism with respect to developing methods,
    - insisting on a priori expectations about the results of analysis,
    - rejection of "strange" results without a further analysis,
    - the idea, that the best way of data treatment is the most comfortable and requiring the least thinking.

Gnostic methodology presents a claim for being a suitable tool for the advanced data analysis.

3.7 The Gnostic Concept

There are three fundamental points differing the mathematical gnostics (MG) from the mathematical statistics:

- Instead of investigating regularities of limitedly large amounts of uncertain events, MG concentrates on uncertainty of one single event, builds its mathematical and physical model and applies it to create the theory of finite (even small) collections of uncertain events.

- Instead of relying on the assumption of existence of a mean and standard deviation of a distribution of a mathematical ideal of a random function,[4] MG takes the uncertain data as images of a really existing events subjected to the laws of Nature and manifesting their real features by their values which are to be respected according to the sound idea "Let data speak for themselves."

---
[4]See the Central Limit Theorem on p. 309 of the book of Mario F. Triola, Elementary Statistics, The Benjamin/Cunning Publishing Co., Redwood City, California

- Instead of mostly relying on the Euclidean geometry and the corresponding Newtonian mechanics, MG applies the Riemannian geometry with Einstein's relativistic mechanics.
- Instead of aggregating the observed data additively, the additive aggregation should be applied to the parameters of the Ideal Gnostic Cycle.

The difference between statistical and gnostic approach to the size of data can be illustrated by Fig. 3.1.

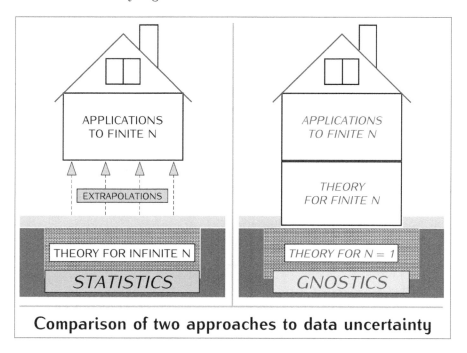

FIGURE 3.1
Statistical and gnostic approach to the size of data

We live in a finite world meeting in our every-day life finite objects and processes. They may be large, but mathematical limits of their amounts and sizes we feel as unnatural and hardly comprehensible. However, statistics builds its theory using infinite limits and satisfies our needs of finiteness by extrapolation from infinity. Unlike this, the finite image of the things is provided by MG by application of the theory of one single uncertain event. To do this, MG uses the four scientific bases (measurement theory, geometry, physics and mathematics) differently. It is illustrated in Fig. 3.2.

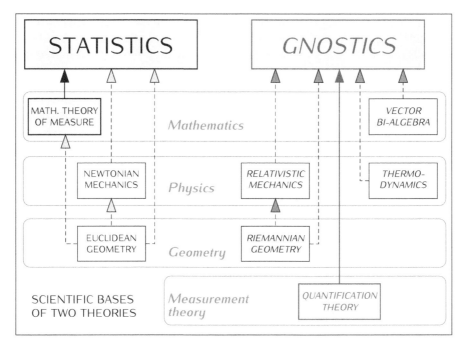

FIGURE 3.2
Main scientific bases of statistical and gnostic approach

3.8 Conclusions

Mathematical gnostics comes with an approach to the variability of (1) data and (2) data analysis, which is as original that represents a change of paradigm. Acceptance of a new paradigm is always difficult and like a change of inertial path of movement of a body it needs an energetic impulse. The role of such an impulse are ordinarily played by both inner conflicts within the valid paradigm and in its applications. Such inner conflicts of the mathematical statistics—the ruling paradigm—originate between its experts and conflicts in applications appear due to serious limitations of the statistical approach. The gnostic paradigm is based on a series of principles enabling to recognize the true causes of uncertainty and the laws to which they are subjected. This enables the uncertainty to be evaluated and their impact on the results of analysis to be diminished. To do that, it needs involvement of unusual steps requiring an increased attention and patience of the reader and user. The basic instrument of statistics—the Law of Large Numbers—can be demonstrated easily by simple tossing a cube or coin, but this is not possible with gnostic paradigm. However, the real power of the new paradigm is clearly testable by using the gnostic algorithms. Unlike statistics determining the weight/importance of an individual data item according to the weight of its family/class, the mathematical gnostics considers the weight of individual data item as a function of its own individual error.

The "cutting the Gordian knot" dreamt up by already cited T. L. Fine consists mainly in three fundamental deviations from the statistical approach: on the concentration on the regularities of an individual uncertain event, on the respect to the data values and what they say about themselves and on consequent use of Riemannian geometry and its application to the recent mechanics. Fine's worries on "where would such a drastic step leave the world of practice" can be considered as void because the applications of mathematical gnostics appear to overcome the statistics in important cases.

4

Quantification

4.1 Ideal Quantification

Natural objects have qualitative and quantitative features. Quantification will be understood in a narrow sense as the act of mapping the real quantities into numbers. In answering the question "how many" it includes *counting* using the integers. To answer the question "how much," the *measuring* is applied, which needs an instrument or measuring device to establish how many times the measured quantity exceeds the *measuring unit*. This mapping is unusual for a mathematician because it is not "from mathematics to mathematics" but "from a real life to mathematics." Everything is precisely defined and processed in mathematics but not in real life. However, the requirements to results of quantification are similar to products of mathematics, especially those relating to consistency of products of quantification (*data*) with the true quantities. These problems led to introducing mathematics to quantification and to the formation of the *measurement theory*, especially after contribution of H. von Helmholtz [105] The real quantities are considered as *mathematical structures*, i. e. as sets provided with a structural operation and satisfying some algebraic relations ([27]). The operation and the relations are defined on real quantities and they may include some people's manipulations. They satisfy some mathematical features and as such they can be mapped as mathematic structures. The operation is thus e. g. mapped as aggregation, the relations as symmetry, coincidence, transitivity, commutativity, reflexivity or irreflexivity, equivalence, and others. This rich offer of relations was summarized in the first gnostic publication [44] as the assumption that the real data are elements of the Abelian group[1]. Particularly, the set of all real numbers with the operation of addition is an Abelian additive group. If a subset of it will be composed of results of quantification, it will be called the *additive data*. The set of data transformed by exponentiation of additive data will be called *multiplicative data* as this will be a subset of an Abelian multiplicative group.

The idea of an ideal quantification is illustrated in Fig. 4.1.

---

[1]The Abelian group is an algebraic structure consisting of a set of elements equipped with an operation that combines any two elements to form a third element and that satisfies four axioms: closure, associativity, commutativity and invertibility.

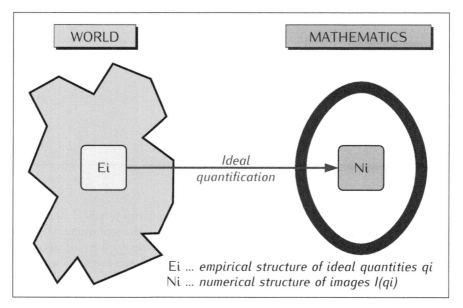

FIGURE 4.1
The ideal quantification

Mathematicians take care to show their realm as an independent world where only the mathematical objects live and laws of mathematics rule over all activities unlike the real word with all its complexities and irregularities. The case of data treatment is a rough exception of this idea. Real data enter the "glass closure" of mathematics as a "foreign" visitor not strictly, mathematically defined. Mathematics doing its best to representing the collection of events that to be mapped into mathematics, considering them as forming some structures known from mathematics, empirical structure E. Mathematical mapping transfers them into mathematics as the numerical structure N.

The problem is that ideal quantification as a real technology does not exist. It is an every-day experience that practical counting and measuring is more or less disturbed by uncertain (usually unexplained) impacts. Consider the simplest case, counting of sheep. It necessarily includes recognition, the success of which depends on many factors. Remember the Homer's Cyclop unable to distinguish Odysseus' soldiers covered by skins from actual sheep because of being blinded. How should a shepherd account for a gravid sheep ready to increase the number of sheep or for a seriously ill sheep which will immediately die? The number of sheep is not a constant, it is always subjected to uncontrolled changes. It is uncertain. Absolutely precise measurements are impossible as well. All this is well known to specialists of measuring but they leave the treatment of uncertainty of measurements to "specialists of uncertainty" for which they ordinarily take statisticians. However, the aim to consider individual uncertainty requires a richer model of quantification.

4.2 Real Quantification

The statistical notion of randomness is agnostic introducing the uncertainty in data values formally, without accepting their natural, material causes. Mathematical definitions of randomness encompass notions of lack of order, periodicity, pattern, aim, or purpose. Unlike this, one can believe that each disturbance of counting and measuring has a natural—may be unknown—cause and as such it could be as measurable as the true quantity, at least theoretically. Let us consider a simple example: the task is to measure the length of a metallic rod. We know, that this length strongly depends on the temperature. The instability of the temperature will disturb the length measurements. The temperature can be also measured like the length of the rod. This allows a choice of two alternatives of the task, to measure the length keeping the temperature as constant as possible or to measure the size at different temperatures. The uncertainty of the rod's length will disturb this measurement. There is no randomness involved, no lack of order, periodicity, pattern, aim, or purpose. The cause is the lack of information on the rod's length. Instead of randomness, we have two cases of ideal quantification, one to quantify the length and one for quantification of the temperature. In a more general case, the real quantification process will include one ideal quantification channel to quantify the true value and the other one to quantify the impacts of uncertainty. They both map empirical Abelian commutative groups of real (true and disturbing) quantities into numbers forming these numeric groups. However, there is an important condition: the operations of both groups are the same. The single number representing the disturbed true quantity is obtained as the operation applied to numeric images of the mapped quantities. The justification of the coincidence of the operation is natural. So for example, if the true value is measured electrically, the disturbing uncertainty will have also an electric impact on the quantifying channel. The idea of the real quantification as a double of ideal quantifications is represented in Fig. 4.2.

There therefore are two empirical structures, Ei of the ideal (true) quantities to be measured and Eu the empirical structure of the uncertain impacts on the measurement. They both have their numeric images brought by ideal quantification into mathematics, but existing only theoretically. However, the structures Ei and Eu interact with each other by entering the structure operation of aggregation the result of which is symbolically denoted as $Ei + Eu$. This is what is really observed and transferred to mathematics. It is then a task for mathematics to treat the "sum" of both structures and to separate them by estimation.

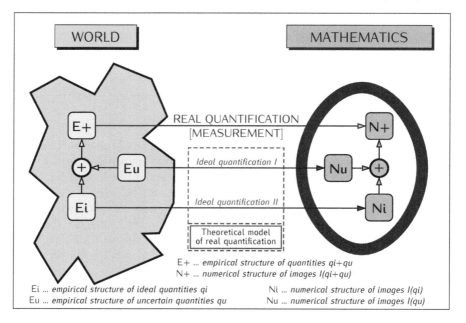

FIGURE 4.2
The real quantification as a double of ideal quantifications

Let us consider the simplified axiom 1 of the gnostic theory 2.2: Let Γ_τ be the empirical additive group of true values and Γ_ν the empirical additive group of uncertainties and $K \cdot A_0$ and $KS\Phi$ their numeric images. Then the uncertain observed data value is obtained as the value of

$$A = K \cdot (A_0 + S\Phi) \tag{4.1}$$

where K is a "physical scale parameter" determined by the physical measuring unit (meter, liter, a.s.o.) and its order of magnitude (kilo-, nano-, a.s.o.). The dimensionless parameter S characterizes the intensity of the uncertainties Φ. To analyze the impact of uncertainty, the additive form (4.1) is transformed to the multiplicative one by exponential transformation:

$$Z = \exp(A/K) = Z_0 \exp(S\Phi) \ . \tag{4.2}$$

Remembering the identity $\exp(S\Phi) \equiv \cosh(S\Phi) + \sinh(S\Phi)$, one can write the multiplicative model of an observed data item as

$$Z = Z_0 \cdot (\cosh(S\Phi) + \sinh(S\Phi)) \tag{4.3}$$

which can be depicted as a point in a plane with coordinates $\cosh(S\Phi)$ and $\sinh(S\Phi)$. If the quantity $Z(S\Phi)$ is a result of observation, we can think of it as passing from the true value Z_0 to the observed point $Z(S\Phi)$ along a continuous path as "driven" by the uncertainty Φ. This is a special path, because for all values of the uncertainty $S\Phi$, equation

$$(\cosh(S\Phi))^2 - (\sinh(S\Phi))^2 = 1 \qquad (4.4)$$

holds. A curve with a constant distance (*radius*) of all its points from a fixed point (*center*) is called a circle. However—as seen from the form (4.4) of this radius, it is a circle in Minkowskian and not Euclidean geometry. There is a substantial difference between both geometries consisting a.o. of an opposite variation principle: the length of a connection between two points by a straight line in Euclidean geometry is a minimum of length of all alternative paths connecting these points. The opposite is true in Minkowskian geometry: the straight line connection has the maximum length among all paths connecting the points. This difference makes the Minkowskian circle an *extremal*, i.e. a curve connecting two points (Z_0 and Z) by an extreme (maximum) path. The proof and details of this statement are in [62].

There is an extremely important consequence of the Minkowskian character of the quantification process: the coordinates of the quantification movement are hyperbolic cosine and sine like at energy and momentum of an unloaded relativistic particle. Therefore an isomorphism exists between the movement of an variable data item under the impact of data changes and a movement of the relativistic particle. Everybody believing in the unity of Nature and interdependence of natural processes should accept this isomorphism doing nothing not respecting it. But mathematical statistics uses an aggregation principle of data which contradicts to this isomorphism because uncertain data are aggregated additively with respect to data values in statistics. The contradiction is in the validity of the Energy and Momentum Conservation Law of the special relativity according to which the energies and momenta of unloaded particles are aggregated additively as hyperbolic cosines and sines of their values, which is nonadditive with respect to their values. The data-additive aggregation is in order in the classical mechanics valid only for the movements much slower then speed of light, therefore for relatively weak uncertainty of data. The im-proper way of aggregation is applied in empirical investigation as well as in preparing statistical measures like mean values, variance, standard deviations, correlations and other. The "named" distribution functions filling the statistical tables are obtained as mean sums of data values thus being in contrast to the laws of relativistic physics.

4.3 Conclusions

The necessity to evaluate (quantify) the quantity of things appeared historically by the progress of people's productivity of work, when it reached such a level, that a man was able to create more things then he consumed. This enabled him to enter the market where he had a chance to change his surplus for another goods. To answer the question "how many" required counting but to determine "how much" it was necessary to develop the technology of measurement based on a measuring unit. The quantification consisted in establishment of the ratio of the quantity to be quantified and the unit. After thousands of years the technology of quantification became to be a matter of scientific efforts by development of the measurement theory. It investigated the individual steps of quantification and established rules for their non-conflicting application. The problem with uncertainty disturbing the process were left to statistics. This is why this way of quantification was called *ideal quantification*. The problem of real-data variability has been included in gnostic approach as *real quantification* consisting of two parallel ideal quantifications of the true value and the disturbing quantity. The rules required by the measurement theory to structure's relations were simplified by the first gnostic axiom requiring the structures of objects to be quantified as the Abelian groups. This model enabled the virtual path of the observed quantity to be defined as an arc of a Minkowskian circle, which is an extremal maximizing the distance of the observed quantity from its true value.

The gnostic model of quantification using the Minkowskian geometry leads to isomorphism between quantification of uncertain data and movement of unloaded relativistic particles. There exists a contradiction of statistical aggregation of real data additive with respect to data values and this isomorphism.

5
Estimation and Ideal Gnostic Cycle

5.1 A Game with Nature

The free traveling along the geodesic (extremal) paths is possible in physics due to balance of powers acting on the really moved mass. But the "movement" of the observed value from the start point (the true value) to the end point (the observed value) along the arc of Minkowskian curve is only virtual. It should help in understanding the process. By introducing the maximum of uncertainty into the quantification process (maximizing the distance/error), the Nature "makes its best" to make the estimation of the true value most difficult. The "counter-move" of the observer who accepts the game should be to look for such an estimation path, which will minimize the uncertainty of the result. To play the game orderly, he applies the science: as proved in [22] there are two gradients of the geodesic in all its points, one directed to maximum growth of its value and another one perpendicular to it, showing the estimating direction of the maximum descent. The latter vector is perpendicular to the circle in the Minkowskian plane, the center of which coincides with that of the quantification path. Following the direction shown by this gradient, one follows the path of the steepest descent. This circle is also an extremal because of connecting differential paths of the minimum length. This circle—already of the "ordinary" (Euclidean) form can be used as the desirable estimation path. The quantification as a mapping from real word to mathematics resulted in the value of exponential function of the real argument $S\Phi$ (4.2) which was naturally decomposed in (4.3) into two additive parts. However, the same observed point can be also depicted as a complex number in mathematics allowing the decomposition into two parts by using the imaginary indeterminate $i = \sqrt{-1}$

$$Z_0 \exp(iS\Phi) \equiv Z_0 \cdot (\cosh(iS\Phi) - i\sinh(iS\Phi)). \qquad (5.1)$$

However, to escape of functions of complex arguments, suitable real functions of the real arguments are to be used to represent the estimation process.

5.2 Double Numbers

The double forms of the observed value was inspired by the features of the exponential function, however there might be a deeper motivation: the notion of "double numbers" has been used for special form of couples of real numbers (a, b) in [35] is written as

$$x = a + c \cdot b \tag{5.2}$$

where c is an indeterminate. It has been proved in the Yaglom's book, that there exist just three fundamentally different versions of c equal to $J = \sqrt{+1}$, $I = \sqrt{-1}$ and $O = \sqrt{0}$. The last case can be considered as a two-dimensional model of Galilean mechanics: take number a as a space and b as time coordinates. Distance between two space-time events $[a_1, b_1]$ and $[a_2, b_2]$ is $\sqrt{(a_1 - a_2)^2 + O^2(b_1 - b_2)^2}$, i.e. the time difference does not play a role at all, because of unwillingly assumed infinite speed of any movement including light. The J-version is interpreted as a point in a Minkowskian plane representing a two-dimensional space-time event of Einsteinian special relativity theory. However, we have obtained two other interpretations connected with the data treatment:

$$Z = Z_0 \cdot (\cosh(CS\Phi) + C \sinh(CS\Phi)) \tag{5.3}$$

where for $C = \sqrt{+1}$ the result of quantification takes place and the complex version with $C = \sqrt{-1}$ could model the estimation but was rejected because of the aim to use real functions of real argument. The task of looking for the parametrization of the *estimating* process required to find such arguments $S\phi$ that the alternative of (4.4) would have the "ordinary" form

$$(\cos(2S\phi))^2 + (\sin(2S\phi))^2 = 1 \tag{5.4}$$

and that the consistence with the quantifying model would be reached. To do this, four data characteristics are to be applied introduced in the chapter Axioms.

5.3 Gnostic Data Characteristics

It may be interesting to see the forms of the quantifying characteristics (Fig. 5.1).

The quantities h_J play the role of gnostic quantifying evaluation of observing error. One would expect their linear or logarithmic model over the axis of multiplicative error Z_i/Z_0, but the actual form of dependencies is different as it strongly dependent on the value of the scale parameter S. The less the S is,

Gnostic Data Characteristics

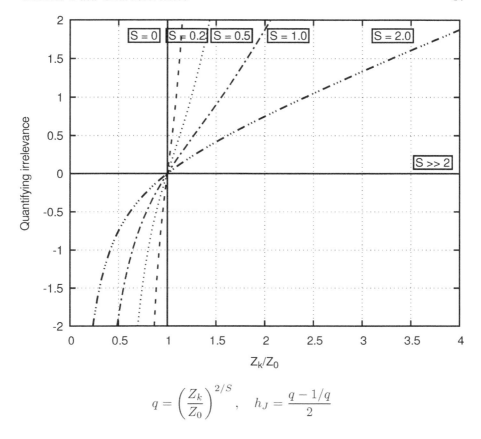

$$q = \left(\frac{Z_k}{Z_0}\right)^{2/S}, \quad h_J = \frac{q - 1/q}{2}$$

FIGURE 5.1
Quantifying irrelevances with different scale parameters S

the steeper the evaluation of the error in dependence on the value of Z_i/Z_0: For $S < 1$ the rise of error evaluation is faster than the rise in Z_i/Z_0. With $S = 0$ the evaluation would be infinity for an arbitrary Z_i different from the Z_0. All this indicates the robustness with respect to the inliers: the larger the distance of Z_i from the Z_0, the larger the error's evaluation. The sensitivity of the evaluation to outliers is emphasized, the inliers are relatively suppressed.

The quantifying variabilities behave in accordance to the quantifying irrelevances, the outlying data are getting the larger weight the farer their value from the sample's center is, the outliers are preferred before inliers. However, the behavior of the estimating weights is opposite: the inlying data are preferred before the outliers by receiving larger weights. The robustness of estimating variables with respect to outliers is thus demonstrated.

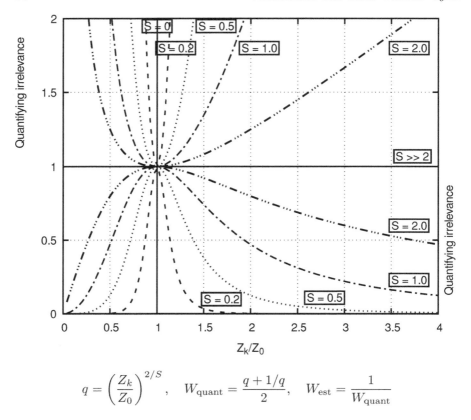

FIGURE 5.2
Quantifying variabilities and estimating variabilities with different scale parameters S

Two relations between both kinds of gnostic parameters of uncertainty immediately follow for a given parameter R defined by (2.9) and (2.15):

$$f_J = 1/f_I = (R + 1/R)/2, \tag{5.5}$$

and by (2.16)

$$\tanh(S\Phi) = \sin(S\phi) = (R - 1/R)/(R + 1/R). \tag{5.6}$$

The former relation reveals the interdependence between quantifying and estimating results while the latter shows the interdependence of the same angles measured by using two geometries, the Minkowskian and Euclidean ones. These relations are important, because

1. The probability of a single data item (will be shown later in chapter Thermodynamics) is

$$p = (1 - \sin(S\phi))/2 \,. \tag{5.7}$$

The Ideal Gnostic Cycle 39

    2. The quantifying probability is not always equal to the estimating one because of a different scale parameter.

5.4 The Ideal Gnostic Cycle

The observed value was 2.4 with the Minkowskian angle $S\Phi$. The same observed value expressed in Euclidean geometry is

$$Z = Z_0 \cdot (\cos(S\phi) + \sin(S\phi)) \tag{5.8}$$

with the Euclidean angle $S\phi$. Both $S\Phi$ and $S\phi$ are known from the observations and estimation. Equation (5.4) holds not only for the angles of the observed value but for all points of the circle. This circle is also an extremal with a minimum length. Imagine now the virtual movement along the Euclidean circle from the (unknown but estimated) observed values $S\Phi$ and $S\phi$ to their mirror images corresponding to the angles $-S\Phi$ and $-S\phi$. The virtual movement along the mirrored halve of the quantification path to the point Z_0 closes the virtual quantification-estimation cycle. This theoretical representation of the processes will be called the Ideal Gnostic Cycle (IGC) and its interesting features will be further investigated.

    A vector connecting a point on a circle with the circle's center is a radius vector. A movement of the circle's point causes a rotation of the radius vector. Two circles and two radius vectors are under consideration:

1. Minkowskian radius vector rotating from the point Z_0 to $S\Phi$ and then from $-S\Phi$ to Z_0.
2. Euclidean radius vector rotating from the point $S\phi$ to $-S\phi$.

Rotating operations in IGC are $S\Phi$, $-2S\phi$ and again $S\Phi$.

    The most important result of the analysis is the difference $f_J - f_I$ which evaluates the overall amount of uncertainty within an IGC. Using relations 2.9, 2.15 and 5.6 one obtains the equality

$$f_J - f_I = h_I/h_J \tag{5.9}$$

enabling to the analyst to "feel" the impact of the IGC because he ordinary knows the relative errors of his calculations. The difference of the quantifying entropy increases and its (partial) estimating compensation by the estimating entropy $RE = f_J - f_I$ deserves a name. It will be called the "residual entropy."

5.5 Information Perpetuum Mobile?

The second law of thermodynamics has several formulations from which that of Sadi Carnot (1824) is probably the oldest: "The natural tendency of the heat is to flow from high temperature reservoir to low temperature reservoir." This statement of fundamental importance is frequently interpreted as a statement of impossibility of a perpetuum mobile, a device capable to produce energy at no input. Its generalization is now known as the second law of thermodynamics. There is a statement connected with the IGC applicable not to the thermal but to the information machines. It says:

The residual entropy of the IGC is always non-negative.

It is easy to show the relation

$$f_J - f_I \geq 0 \qquad (5.10)$$

recalling that f_J is a hyperbolic cosine and f_I a trigonometric cosine. The importance of this statement for praxis of analysis of uncertain data is in impossibility of a perfect estimation: the uncertainty in once contaminated data cannot be never perfectly eliminated. Achieving an "information perpetuum mobile" capable of doing this is impossible.

5.6 Existence and Uniqueness of the Ideal Gnostic Cycle

The notion of the *IGC* along with four characteristic functions of uncertain data may seem to be based more on the scientific fantasy than on something real. However, a function *Gmean* exists, described in more detail in chapter *Robust estimation of a constant* enables to present *IGC* as an existing and unique object endowed with extreme relations to the gnostic characteristics.

TABLE 5.1
Three phases of looking for extremes of Ideal Gnostic Cycle

Iteration	Mean(z_0)	Arg.extremity	$R = Z/z_0$	Extreme	Extreme value
1-start	60.42857	61.45898	0.983234	min(f_J)	1.0104620
10-finished	60.42857	59.82464	1.010095	min(f_J)	1.0100950
1-start	60.42857	61.45898	0.983234	max(f_I)	0.9897804
10-finished	60.42857	59.72436	1.011791	max(f_I)	0.9901705
1-start	60.42857	61.45898	0.983234	min($f_J - f_I$)	0.02068166
10-finished	60.42857	59.77595	1.010918	min($f_J - f_I$)	0.01992528

It is an iterating algorithm operating in three phases solving the equation of the data model (2.4) in the form of (2.8). The data sample is column 1 (Air Flow) of matrix *stackloss*.

The IGC thus exists and is unique because everybody can calculate its extreme values in this way in application to his sample of uncertain data. Its objective existence classifies it to the category of Law of Nature which existed always waiting for its discovery.

5.7 Conclusions

The uncertainty of quantification does not allow precise value of the true quantity to be evaluated. This is why it must be estimated. The task of estimation is made harder by the Nature's way of introducing the variability by changes of true values as well as by uncertainty into the quantification to maximize the error of possible estimation. To play this "game" with Nature, the observer is looking for the best (shortest) estimating path from the observed value to the true value. The analysis shows, that such path exists leading along the Euclidean circle to the mirror point of the observed one. The closed path including the Minkowskian arc from the true to the observed value, the Euclidean arc from the observed to mirrored value and closed by another Minkowskian arc from the mirrored to true value forms the IGC. Four gnostic data characteristics identified as parameters of the *IGC* demonstrate different kinds of robustness.

The difference between the quantifying uncertainty and estimating uncertainty is the residual entropy which quantifies the amount of uncertainty of one pass of the IGC. It is always non-negative.

6
Geometry

6.1 A Historical Dispute on Robustness of Statistics

This is a basic and most frequently asked task of statistics: Given a data sample D obtained by N repeated observations performed under "the same" (comparative) conditions. It is required to estimate a single number representing the sample's data which differ by contribution of uncertainty. Such a value characterizes the location of the data on the data domain and is called the *location parameter*. Experience has shown the sensitivity of the classical location parameters (mean, median, distribution mode) to the "bad" data values (e. g. outliers) and the demand for robust methods appeared. Many new methods were developed and tested. A historical case is worth of being remembered [84].

Professor Stigler from the Massachusetts Institute of Technology published paper [34] where apart from two classical methods (mean and median) nine robust and adaptive methods recommended by experts were tested in application to real data. Results of famous physical experiments from the 18th and 19th century were used as undoubtedly real data were selected. The data were organized in 16 samples of about 20 measurements of the same variable from the following experiments:

1. Short's determination of the Sun parallax (1763, 8 samples),
2. Newcomb's measurements of the time of passing a distance by a ray of light (1882, 3 samples),
3. Michelson's measurements of speed of light (1879, 5 samples).

Stigler's results were extended by the twelve method of measuring the location parameter as the mode of gnostic global distribution function *EGDF*. Following criterion functions were used for evaluation of results like in Stigler's study: Let $E_{0,n}$ $(n = 1, \ldots, N_m)$ be the true result of the measurement of the nth variable and $E_{m,n}$ its estimate $(m = 1, \ldots, 12)$ obtained by the mth method. Denote

$$\bar{D}_n = \frac{1}{N_m} \sum_{m=1}^{N_m} |E_{m,n} - E_{0,n}| \qquad (6.1)$$

the mean value of D_n. Thus

$$e_{m,n} = |E_{m,n} - \overline{E_{0,n}}|/\bar{D}_n \qquad (6.2)$$

is a relative error of the nth estimate of the mth variable which can be used for comparison of measuring different variables. Professor Stigler has used in his comparison the recent values of the measured quantity and came to a general conclusion that "the modern robust estimates are not worth of the time of computers to evaluate them." This paper generated a wave of publications discussing the problem. The methodological flaw of Stigler's evaluation was attacked by paper [25] pointing out that the original results were strongly biased. However, the authors of [84] added to methods to be compared the gnostic estimate obtained as the mode of the gnostic probability density function. They took note of the distribution of results of all 12 tested methods and tested it to show its Gaussian character. Considering the account that each of methods was made by an expert who is unbiased, and uses all his rich experience to analyze the data, they used the means of all methods as the true values of the historical measurements: if the methods were known at the time of experiments, this method could be used to evaluate the results. The relative true value of all measurements was put equaling 1. Three methods of comparison were therefore applied:

The absolute deviation $|\mu - 1|$ of the mean error from 1, where

$$\mu_m = \sum_{n=1}^{N_m} e_{m,n} . \qquad (6.3)$$

Mean squared error:

$$\varsigma_m = \sqrt{(\sum_{n-1}^{N_m}(e_{m,n} - \mu_m)^2/N_m)} . \qquad (6.4)$$

Errors range:

$$R_m = \max_n(e_{m,n}) - \min_n(e_{m,n}) . \qquad (6.5)$$

Results of this comparison are in Tab. 6.1.

The positiveness of these results obtained in application to real data motivated several series of studies with simulated data. After the applications to several types of distribution functions from Gaussian through Cauchyan, the data of one distribution were contaminated by data from another distribution. When the contaminating distribution had a symmetrical density, the gnostic and best statistical results differed only by small percentages. However, for non-symmetrical disturbances the error of statistical results was rising monotonously in proportion to the deviation from symmetry, while gnostic results stayed to be usable.

TABLE 6.1
Comparison of 12 methods of estimating the location parameter

m	Type of the estimate	ς_m	$|\mu_m - 1|$	R_m
1	Gnostic	0.038	0.001	0.139
2	Hogg's T1	0.061	0.017	0.261
3	25% trimmed	0.070	0.029	0.261
4	Edgeworth's	0.079	0.011	0.273
5	15% trimmed	0.104	0.032	0.447
6	Tukey's 'Biweight'	0.131	0.043	0.631
7	Andrew's AMT	0.147	0.025	0.660
8	Huber's P15	0.210	0.083	0.856
9	10% trimmed	0.211	0.097	0.821
10	Arithmetic mean	0.212	0.078	1.055
11	Median	0.278	0.124	0.962
12	Outmean	0.619	0.086	2.603

6.2 Distance as a Problem

Data are treated to estimate their true values masked by uncertainties. Ignoring the extremely rare case of precise measurement or counting, the observed value is always different from the true value, and an error takes place. To quantify the error, one uses either the additive form (difference between the true value and its estimate) or the multiplicative form (the ratio of both values). In both cases, one has to apply geometry to evaluate the distance caused by the error. Let us follow [29] to analyze this problem:

Let us consider two points a and b on a straight line with coordinates c_a and c_b, and ask a man in the street the question, "What is the distance (L) between the points?" The answer depends on the level of the reader's mathematical skill:

Basic: It is simple,
$$L = |c_a - c_b| \, . \tag{6.6}$$

Thoughtful: It is not that elementary, because the path of integration, along which the distance should be measured, has not been defined. For the path denoted $\mathcal{P}(a,b)$ the distance would equal to the path integral

$$L = \int_{\mathcal{P}(a,b)} dp \tag{6.7}$$

where dp is the length of an element of the path. Only in the case of the integration, path coinciding with the straight line does the expression (6.6) reduces to the ordinary integral

$$L = \left| \int_{c_a}^{c_b} dx \right| \qquad (6.8)$$

which provides the same result as (6.6).

Advanced: It is complicated because neither the integration path nor the geometry is specified. Assume, that the integration path is $\mathcal{P}(a,b)$, and that such geometry is chosen, that the weight of an element dp of the path of the point z is $g(z)$. Then the distance is

$$L = \int_{\mathcal{P}(a,b)} g(z) dp \ . \qquad (6.9)$$

This expression reduces to (6.7) only if the weight $g(z)$ is a constant equal to 1, which is the case, when Euclidean geometry is employed, otherwise the geometry is of Riemannian type.

The importance of knowing the path, when measuring some distances, can be illustrated by the following: imagine a lady, whose home is at point a and has an office at point b. Even in the case, when both points are on the same straight street, the distance to be actually walked or driven between them would seldom be the same each time. It depends on stops made to shop, visit a hairstylist or a friend, etc. on the way. This problem of the path is important particularly, when measuring uncertainty, because uncertainty can move the geometrical image of data along a curve and not along a straight line.

The need to use a variable weight $g(z)$ in (6.9) can be also illustrated by an example. Consider the process of estimating the value of an asset by several differently qualified experts. Assume, that an "ideal" value for the property exists, e. g. as estimated by an omniscient Expert, but no such expert is on hand. It is felt, that all of these experts, together, should come close to the ideal value. The estimates are real numbers and the evaluation process is begun by calculating their arithmetic average. A second approximation is made to produce a more realistic result; it represents a **weighted** average using weights, which are dependent on the distance of each individual's estimate from the previous round's mean. Weights equal to 1 are given to estimates, which have the value of the previous average and these weights decrease with increasing distances from the mean. The new value of the weighted average is used for the next round and the iterative process ends, when the weighted average remains constant. The estimates of "bad" experts are thus suppressed and those of "good" ones are emphasized. The idea is simple—instead of assuming, that judgments of all experts are the same, and that each have the right to be taken equally into account, we evaluate the individual qualification of each expert by the quality of his estimate. This is accomplished by using a particular weighting function $g(z)$.

There are at least three classes of statistical approaches to the problem of the weighing functions:

- All data are given the constant weight determined by the reciprocal sum of the of data divided by the data number. This is the mostly applied operation of the statistical mean value used to evaluate data means, variances, standard deviations, correlations and other statistics.

- Estimation theorem for normally distributed sub-samples of data with unequal variances: As proved in already cited Linnik's book, asymptotically efficient and unbiased estimate of the common mean value can be obtained by using the data weighing functions equal to reciprocal variance. The application range of this approach is limited by the assumption of data normality and availability of the variance of the sub-samples.

- The Iterated Weighted Least Squares Method based on influence functions [95]: a robust multidimensional regression model can be obtained by application of influence functions to weigh each equation of an equation system while iteratively looking for the solution of the system. Many versions of influence functions are available in the literature robustly working in application to some data while failing on others.

However, a note could close this subsection: Four scientific revolutions seem to be ignored by statistics as a science:

1. Progress in thermodynamics and especially by R. J. E. Clausius' introduction of its conception of entropy.
2. Contribution of Georg Friedrich Bernhard Riemann to the geometry at middle of 19th century.
3. The scientific conception introduced by H. Helmholtz into the theory of measurements.
4. Albert Einstein's generalization of classical mechanics at the beginning of the 20th century.

The founders of statistics were thinking in a "Euclidean-Newtonian" way of their time. There is a question if the way of thinking on statistical problems should not be changed to the "Riemannian-Einsteinian" way?

6.3 Additivity in Data Aggregation

Most of statistical methods are using simple additive aggregation law with the constant weight in application to data values, their squares, products and statistical moments without bothering on justification of this manipulations. Such an additive data aggregation law is being applied to estimate empirical statistics like means, variances and correlation coefficients and other as a primitive way not deserving a scientific justification ("... everybody knows why it should be so ..."). The English units of length (inch, foot, yard, mile) reveal the historical roots of this "natural and self-determined" notion:

Distance x steps, distance y steps $\implies$ distance $x \& y = x + y$ steps.

However, there are ways to further support this manipulation:

1. The quantity of things is expressed in numbers which are members of additive structures like groups, algebras and others.
2. There exists an inter-science isomorphic mapping between statistical variables and elements of Galileo-Newtonian mechanics:

   - statistical variable $\iff$ speed of a body in mechanics,
   - square of the variable $\iff$ kinetic energy of the body,
   - product of two variables $\iff$ turning momentum of a body,
   - the aggregation law in classical mechanics is additive by the Energy and Momentum Conservation Law of classical mechanics. To preserve the mapping, one should accept the additivity for statistics as well.
   - Accepting the Unity of Nature principle, one can see the "natural" additivity of statistics is supported by the Classical Mechanics, which is suitable for the sufficiently slow processes.
   - The founders of statistics were educated in classical mechanics and therefore they probably felt this way of data aggregation as adequate and justified.

6.3.1 Statistical Mean Value and Data Weighting

The statistical mean value defined as the first statistical moment is estimated by the arithmetical mean of data values. The following objections should be considered:

1. The errors of individual data can be different but the weight of data is the same independently of their uncertainty relying on compensation of gross errors by a sufficient amount of small errors of other data.
2. Large individual contributions to estimate's value are accounted for like the small ones.
3. The mutual compensation of errors works according to the Law of Large Numbers under assumptions of a sufficiently large data sets of limitedly uncertain data and the existence of integrals defining the statistical moments. It is difficult or even impossible to verify these conditions for a given data set having an unknown distribution function.

4. Moreover, as follows from mathematical gnostics, the space of uncertain data is curved by the data uncertainty making the relation between data values and their errors non-linear and requiring the non-linear data aggregation law. The effects of space curvature increase with amount of uncertainty. The errors caused by ignoring the curvature also rise with uncertainty.

5. Using the operation *mean*, one gives to all data the same weight independent on their uncertainty. Contribution to the estimate's value is thus following the principle of "collective fault."

6. When the weight given to individual data is determined as reciprocal variance as in already mentioned Linnik's book [101], then the "penalization" of an individual datum by lowering the weight of its contribution is directed also by a collective fault, by the variance of the sub-sample to which the data item belongs.

7. The development of the robust statistical theory brought aims to individualization of weights dependent on the data item uncertainty. The idea of individual weights dependent on the data item uncertainty led the robust statisticians to invention of the *IWLS* method (Iterative Weighted Least Squares) of multidimensional robust regression modeling. However, the data are weighted by the "collective" error of the equation.

The right way of aggregation is giving the individual weight of each data item decreasing by the item's individual "fault"—its uncertainty. The gnostic quantification data uncertainty is individual. It rises with data uncertainty. The estimation certainty falls with uncertainty, their difference (residual thermodynamic change REC, the entropy change) is increasing with the data uncertainty. Looking for minimum of the REC's sum, one ensures the minimum of summary uncertainty of individual data uncertainty. This thermodynamic interpretation of uncertainty enables interesting and useful estimation methods to be developed.

6.4 Double Robustness

Really observed data are finite. Therefore finite values $Z_L(S\Phi_L)$ (the least) and $Z_U(S\Phi_U)$ (the largest one) exist for each observed data sample. Let us now consider the behavior of gnostic data characteristics under changes in their additive parameter $S\Phi$. Data $Z \leq Z_L$ and $Z \geq Z_U$ are ordinarily considered as *outliers*, as bad data. A large effort has been applied by robust statisticians to develop methods preventing the results of data treatment methods against impacts of outliers. The experience of treating real data from different application fields is conforming the importance of this type of robustness. However, there are practical tasks requiring the opposite kind of robustness, the robustness against *inliers*, i.e. against data $Z \geq Z_L$ and $Z \leq Z_U$. Consider examples of such tasks:

1. Astronomy is observing signals from cosmos to look for foreign civilizations. Ordinary response is a noise of limited signals from the interval $[Z_L, Z_U]$. The desirable signals are outliers.
2. The quality assessment control systems are designed to check appearance of products, in which the parameters are going outside the tolerated bounds. The good products (inliers) make "noise" of measurements. The system must be activated by the outliers while passing through the good products.
3. Diagnostic decisions in medicine is mostly based on checking of medical parameters appearing outside the interval of the reference values. The "normal" values are considered as a measuring noise not requiring an intervention of medical persons unlike values crossing over the reference bounds.
4. The environmental parameters are monitored if their values are kept under a limit. Crossing up a limit must be sensitively signaled as an alarm.
5. Capital reserves of a financial institution must be held over a level ensuring the stability. Falling down under the safe level is a matter of increased interest requiring some actions.

In all these and similar cases the methods robust with respect to inlying noise and giving preference to the outliers are required.

Both kinds of robustness are offered by gnostic characteristics of data uncertainty: the estimating certainties prefer the inner data of the sample by giving them larger weights before the outliers obtaining the less weight the greater their distance from the true value's estimate. The opposite effect takes place in the case of quantifying uncertainties. Two kinds of robustness in evaluating the data errors can be seen at irrelevance and relevance as well. However, a third kind of robustness exists as well: the residual entropy, the difference between the quantification and estimation entropies. It combines both "inner" and "outer" types of robustness and may be useful as the criterion for optimization. The entropies connected with gnostic data characteristics will be analyzed in the following.

6.5 The Curvature of the Space of Uncertain Data

Return to (5.1) valid for both verses of the indeterminate c. Interpret it as the point Z in a Minkowskian plane with coordinates $\cosh(S\Phi)$ and $\sinh(S\Phi)$ or

Three Geometries 51

in an Euclidean plane with coordinates $\cos(S\phi)$ and $i\sin(S\phi)$. Let us consider the differentials

$$\cosh(S\Phi) = \sinh(S\Phi)\, dS\Phi \qquad (6.10)$$

and

$$\sinh(S\Phi) = \cosh(S\Phi)\, dS\Phi \qquad (6.11)$$

demonstrating that all differentials of $d(S\Phi)$ are accounting for by integration of the path with the weight $\sinh(S\Phi)$ ($\cosh(S\Phi)$) and not with a constant weight as in the Euclidean geometry. It means, that the geometry is of Riemannian type. It also means that when taking the weights $\cosh(S\Phi)$ and $\sinh(S\Phi)$ playing the role of *metric tensors* as functions of scale parameter S, one sees, that the space is curved, because to evaluate the curvature's radius one would twice differentiate the metric tensor by the variable $S\varphi$ to not receive a zero. It also means that the local curvature's radius will depend on the scale parameter. Analogous consideration might relate to the case of estimating variables.

The scale parameter S (and with it the space curvature) may be independent on the point of the space (on the local uncertainty) or may depend on the local uncertainty. The data with a constant scale parameter will be called *homoscedastic data* like in statistics. The data with a scale parameter dependent on the local uncertainty will be called *heteroscedastic data*.

6.6 Three Geometries

The consideration of the Ideal Gnostic Cycle led to two geometries, the quantifying, Minkowskian (which will be denoted 'Q') and estimating, Euclidean (denoted by 'E'). Two important characteristics of uncertainty were introduced, the quantifying entropy increase f_J and estimating entropy drop f_I also called data weights. But a special law has been shown to be valid: the difference's $f_J - f_I$ (residual entropy) the positivity of which has the power of a natural law like the second law of thermodynamics. Both addends f_J and f_I give rise to a geometry playing the roles of metric tensors. They are additive and their difference can thus be also used as a metric tensor to create a third geometry for the data treatment which will be denoted by 'R'.

The entropy f_J is a hyperbolic and f_I trigonometric cosine. Taylor's polynomial of their difference is

$$f_J - f_I = 2(x^2/2! + x^4/4! + x^6/6! + \ldots) \qquad (6.12)$$

It may be interesting to see the forms of the gnostic characteristics.

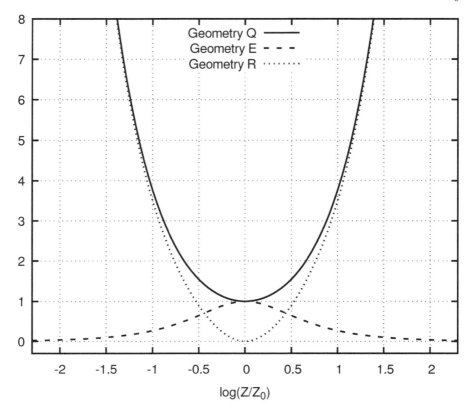

FIGURE 6.1
Three gnostic data characteristics, f_J, f_I and $f_J - f_I$ (geometries 'Q', 'E', and 'R')

6.7 Conclusions

An error is a distance from the true to estimated value. As such it is an object of geometry. There are a number of geometries, each of which determines its specific way of evaluating the distance. Establishing the metric of the space of uncertain data is therefore a non-trivial problem. The processes of quantification and estimation are followed by using two different pairs of coordinates called data certainty and uncertainty and data irrelevance and data relevance (or gnostic data characteristics). Their analysis shows, that both data certainty and uncertainty converge with decreasing amount of uncertainty to statistical variance while the irrelevance and relevance converge to statistical additive data errors. It means that they may be considered as generalization of statistical notions of variance and data errors. The behavior of gnostic data characteristics under changes of uncertainty amount is opposite. They are not quadratic like variance and they are not linear like the statistical errors. The quantifying characteristics prefer the outlying data before the inliers by giving them larger weights. Contrary to this, the estimating characteristics prefer the inlying data before the outliers. In this way, two classes of robustness with respect to data are achieved by two classes of gnostic characteristics. Another important feature of the characteristics is its dependence on the scale parameter S. Its value determines the local curvature of the space of uncertain data.

The difference of the quantifying and estimating entropies (the residual entropy of the Ideal Gnostic Cycle) applied as a metric tensor gives rise to the third geometry for data treatment.

Estimation of a mean value is a most frequently asked task of data treatment: to represent an one-dimensional data sample by a single number. There are two key problems in solving this task, the weight of each individual data item and the way of aggregating the data weights of all data.

The uniform weight applied by calculating the statistical mean value suffers from the non-robustness caused by strong differences between data values. Application of the weights proportional to the reciprocal variances of the data sub-samples requires availability of cluster analysis of data sample and does not ensure maximal information of the results, because the weights given to individual data items do not correspond to the item's own uncertainty but to the "collective" uncertainty of the sub-sample. As results from the mathematical gnostics, the maximum of result's information is reached by data weighing by individual weights dependent on the individual uncertainty of the data item.

It also results from the gnostic theory that the additive aggregation of the data and their powers and products applied in statistics is applicable only in cases of a weak data uncertainty. The right and more universal way of data aggregation is additive aggregation of gnostic characteristics of data which are non-linear data functions.

7
Aggregation

The term *aggregation* is used in abstract algebra for a binary operation creating a new element of a structure by the composition of two structure's elements with the accumulation of their values or of their functions. An example may be the additive addition of real numbers or the multiplication of positive real numbers. However, the way of aggregation may be very different from addition or multiplication of numbers. So, e.g. in structures of living organisms, the aggregation includes sexual relations. It is therefore a question, how characteristics of uncertain data are aggregated. The first gnostic axiom requires for the aggregation of data additively or multiplicatively. But does it mean, that characteristics of data uncertainty should be always aggregated additively or multiplicatively as well?[1] This question is worth of investigation because it is not a trivial matter.

Axiom 1 was applied to definition of two data models and four data characteristics, $\cosh S\Phi$, $\sinh(S\Phi)$, $\cos S\phi$ and $\sin(S\Phi)$. Four ways of aggregating them additively can thus be justified for each of the six objects:

1. All of them are represented as real or positive real numbers, members of the additive or multiplicative algebra the structure operation of which is commutative.

2. Additivity of measured data applied in statistics is therefore natural.

3. Each of the hyperbolic and trigonometric data characteristics are functions of an argument (Minkowskian or/and Euclidean angle) which can be aggregated additively.

4. Additivity applied to each of four data characteristics results a.o. from their membership in commutative algebras.

Different ways of aggregation lead to different results. Which and why should be used in applications? The decision was done by Axiom 2, which is supported by the Laws of mechanics.

---

[1] Additivity of multiplicative data is implemented via the commutativity of the structure operation

7.1 Why the Least Squares Method (Frequently) Works

It is useful to borrow the notion of the *transformation* from H. S. M. Coxeter [32]: "We shall find it convenient to use the word transformation in the special sense of a one-to-one correspondence $P \to P'$ among all points in the plane (or in space), that is, a rule for associating pairs of points, with the understanding that each pair has a first member P and a second member P' and that every point occurs as the first member of just one pair and also as the second member of just one pair..." If this transformation is *one-to-one*, the transformation is called *isomorphism*.

Let us consider a finite structure $\mathcal{D}$ of the sufficiently precise uncertain data d and a structure $\mathcal{M}$ of load-free particles of mass m moving with such a sufficiently low velocity v, that the Galilean/Newtonian mechanics provides an acceptable description of their movement. The number/quantity of members of both structures is L. Data uncertainty of the i-th data item d_i is characterized by a pair $D_i \equiv [e_i, e_i^2]$. In a Gedanken-experiment we can think of the particle m_i corresponding to the data item d_i. Its dynamic state is characterized by the classical momentum-energy pair $M_i \equiv [m_i v_i, m_i v_i^2/2]$. Establish an *interscience isomorphism $\mathcal{ISMA}$* by an one-to-one mapping the elements $d_i \leftrightarrow m_i$ for all i.

The aggregation law of particles is given, it is the additive momentum-energy law of classical mechanics. To preserve the isomorphism $\mathcal{ISMA}$ of structures' elements, one must introduce the additive aggregation to the pairs D as well. It means that additive aggregation law for data errors and their squares is supported by the natural Momentum-Energy Conservation Law of classical mechanics.

The first line in Fig. 7.1 shows the aggregation of classical mechanic momenta which are aggregated additively by the Momentum-Energy Conservation Law of classical mechanics, which is represented as a hyperbolic sinus like the quantifying irrelevance of an uncertain data item in a case of weak uncertainty. An isomorphism M can be established between uncertainty and mechanic momentum. The momenta are aggregated additively. The irrelevances should be also aggregated additively to let the isomorphism hold for aggregated items as well.

The analogous diagram can be drawn for an isomorphism between the data squares and energy of a particle moving sufficiently slowly to be aggregated additively by the Momentum-Energy Conservation Law of classical mechanics. To close the diagram, additive aggregation law should be accepted for squares of a weakly uncertain data items.

Many people believe that the popularity of the Least Squares Method (*LSM*) applied to statistics results from the fact, that to minimize the least squares one has to solve only a system of linear equations by methods known and easily available even before the computer times. But the experience with

Aggregation of Uncertain Data 57

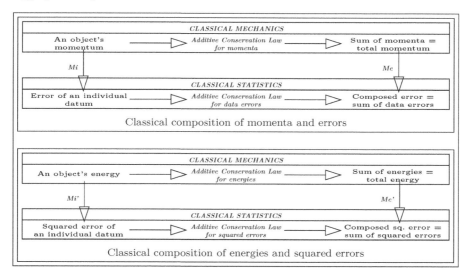

FIGURE 7.1
Diagram of mapping in the classical aggregation

LSM was surprisingly good, it worked even in tasks, where other methods were failing. In the light of the just done analysis it is possible to state:

1. The good results of the *LSM* resulted from its correspondence with the natural Momentum-Energy Conservation Law of classical mechanics.
2. The cases of failing of the *LSM* can be caused by the data uncertainty crossing the bound of sufficiently small uncertainty.

The accepting of the inter-science isomorphism is based on the idea of the Nature's unity: some fundamental rules are valid in different areas of life. It cannot be excluded that the author of *LSM*, Carl Friedrich Gauss, was motivated in discovering the method in 1795 by his knowledge of mechanics of celestial bodies by the idea, that the quadratic distance of the observed location from the inertial path should be used as the weight of observation. The energy necessary to pass such a distance was proportional to squared distance and the momentum corresponded to errors.

7.2 Aggregation of Uncertain Data

To consider the aggregation of uncertain data more generally not limiting the amount of uncertainty, one has to take in account four facts:

1. Data are subjected to Minkowskian geometry during quantification.
2. The classical mechanics is valid only for small velocities.
3. The isomorphism $\mathcal{ISMA}$ was introduced only for sufficiently precise data.
4. The data characteristics are functions of only one variable, $S\Phi$ (or $S\phi$).

Actually, the pair $[f_J, h_J]$ can be rewritten as in Lorentz transformations

$$f_J = 1/\sqrt{1 - (h_J/f_J)^2} \tag{7.1}$$

and

$$h_J = (h_J/f_J)/\sqrt{1 - (h_J/f_J)^2} \,. \tag{7.2}$$

Compare these relations to formulae of components of bi-dimensional tensor of energy and momentum of a flow of density of mass [28]

$$QW = 1/\sqrt{1 - (v/c)^2} \tag{7.3}$$

and

$$QH = (v/c)/\sqrt{1 - (v/c)^2} \tag{7.4}$$

where v is the relative velocity of the observer with respect to the moved object and c is the speed of light. It is obvious, that an isomorphism $\mathcal{ISMB}$ exists of the form $h_J/f_J \leftrightarrow v/c$ between an uncertain data item and an object of the relativistic mechanics. However, the aggregation law for the relativistic pair exists, which is the Energy-momentum Conservation Law of Einsteinian mechanics and is additive. To keep the isomorphism $\mathcal{ISMB}$ valid, one has to accept the additive aggregation law for gnostic characteristics h_J and f_j as well. It is easy to see that isomorphism $\mathcal{ISMB}$ is—as physicists say—Lorentz invariant, i.e. it holds for all values of velocity and amounts of uncertainty changed by Lorentz's transformation.

A diagram in Fig. 7.2 analogous to that of the classical mechanics illustrates this principle.

The name of the variable *gnostic weight* in the figure should be read as quantifying data uncertainty. The uncertainty actually plays role similar to that of the mass multiplying the energy and impulse of the relativistic particle. But uncertainty of a data item is dependent on the value of the scale parameter. The scale parameters of different data items may differ in case of heteroscedastic data. It means, that individuality of data scale parameters must be accounted for in data aggregation. Some lessons for the mathematical statistics result:

1. The data relevances and irrelevances, certainties and uncertainties of uncertain data that are not sufficiently precise should be aggregated additively.

The Second Axiom 59

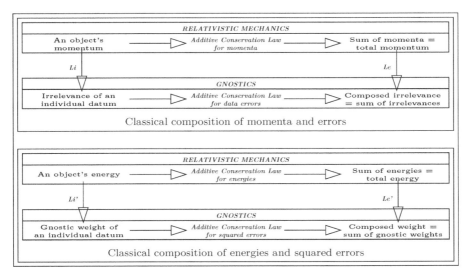

FIGURE 7.2
Diagram of mapping in the relativistic aggregation

2. Such data itself and their squares should not be aggregated additively because their irrelevances and relevancies, certainties and uncertainties are non-linear functions of data values.

3. Estimation of scale parameters of data to be aggregated as well as the decision if the data are homo- or heteroscedastic is necessary.

4. Not only estimation of statistics but all algebra of uncertain vectors and matrices should be adapted to the proper aggregation law. The simple notion of the *mean value* is becoming to be more complicated requiring some data analysis.

We have already seen that additive aggregation law applied to sufficiently precise data and their uncertainty is legal because of correspondence to classical mechanics. It can be now added, that additive aggregation is acceptable *only* to sufficiently precise data.

7.3 The Second Axiom

The validity of foregoing consideration of aggregation is dependent on the believe in unity of Nature reflected by an universality of rules valid in different areas of life. Respecting the character of mathematical gnostics as an axiomatic theory, one has to accept the validity of the isomorphism $\mathcal{ISMB}$ as the second axiom of the theory with an extension: each quantification event characterized by the pair of characteristics $[f_J, h_J]$ is accompanied by an estimation event $[f_I, h_I]$. There is a relation (5.9) bounding both kinds of pairs together, so there is an one-to-one mapping between both kinds of characteristics. This leads to inclusion of acceptance of the additive aggregation of estimating data uncertainty and relevance under the power of the second gnostic axiom.

7.4 Conclusions

The frequently observed success of the Least Squares Method can be explained by existence of a mapping between the pairs of data squares and linear data errors and pairs of the energy and momentum of classical mechanics. Additive way of aggregation of data squares and data values adopted in statistics is thus supported by the Energy-momentum Conservation Law if the data are sufficiently precise. Quantifying data weights and irrelevances applicable to more general cases of uncertainty can be shown to be in isomorphic relation with the relativistic energy-momentum pairs. It means that the additive aggregation of quantifying weights and irrelevances is supported by the Energy-momentum Conservation Law of relativist mechanics. It also results that the uncertain data and their squares not being sufficiently precise should not be added additively. To keep the character of mathematical gnostics axiomatic, this statement is considered as the second axiom with its extension valid for the estimation uncertainties and relevances.

8
Thermodynamics of Uncertain Data

A noteworthy expression of A. Einstein related to the thermodynamics is as follows:

Thermodynamics is the only physical theory of a general nature of which I am convinced that it will never be overthrown.

The thermodynamic variables are of deterministic nature; they are observable and measurable, but with one exception—entropy. The entropy is one of the most mysterious natural quantity defined by Rudolf Clausius in 1855 as the ratio of heat to absolute temperature. It is neither observable nor directly measurable, but so powerful, that its participation in all natural processes deserved the formulation as the Second Law of Thermodynamics. The significance of this original notion has been evaluated in [104] in this way:

As shorthand for the useful but meaningless ratio of heat to temperature it continues to be part of the jargon of chemistry and engineering. Later, when it was finally revealed as a measure of molecular disorder, and of the amount of wasted energy in heat engines and other thermal processes, it would rise to a position rivaling that of energy in our understanding of the universe.

Ludwig Boltzmann redefined the entropy for a system of molecules of the ideal gas. His formula was taken over by Claude Shannon in 1948 as a measure of disorder of uncertainty evaluating the amount of information. Both formulae require a complete availability of all the probabilities of events to be considered. Neither Boltzmann nor Shannon bothered to show that their entropy is actually the same notion as that of Clausius's neither that their formulae are valid even for individual uncertain data. This task was left for the mathematical gnostics which has built a complete model of data uncertainty and related information on the thermodynamic entropy of Clausius's kind.

8.1 Thermodynamic Interpretation of Gnostic Data Characteristics

Let us return to showing the observed value as two versions of double numbers $x + cy$ in dependence on the value of the indeterminate c ($\sqrt{(-1)}^2 = 1$ or

$\sqrt{(-1)^2} = -1$). A Gedanken-experiment could follow. Data are quantified mostly by electronic devices. In cases when quantification is not electronic, one can imagine the electronic model of the process, e.g. by means of an analog computer. It is therefore possible to accept the idea, that the result of quantification is represented by an electrical quantity, voltage, current, number of impulses a.s.o. In all such cases the energy of the image of the measured quantity will be proportional to the square of its value. Consider a computer screen, the neutral $(x+cy) \equiv (0+c0)$ electron beam is directed to the screen's center while the horizontal deviation is controlled by x and vertical by cy. The energies of both deviations will be x^2 and c^2y^2 their sums equalling $x^2 + c^2y^2$. The energy can be converted into temperature e.g. by heating or cooling a material object (in dependence on the sign of c^2). The object's temperature T will be proportional to the energy. The transferred amount of heat Q could be measured by a calorimeter. This can be done for a value of c and for its opposite value c' as well. Denoting e_J the quantification energy ($c^2 = 1$) and e_I the estimation energy ($c'^2 = -1$), one can follow the energy difference $e_J - e_I$ during both processes of passing both paths. Energy e_J during quantification rises due to the contribution of uncertainty and e_I stays to be constant. Energy e_J is constant during estimation but the e_I is increasing. The energy difference of both phases and the amount of an element of heat is thus $(k_Q 2d(cy)^2) \cdot (e_J - e_I)$. The temperature during both phases is constant, equalling $T = k_T(x^2 - (cy)^2)$. The Clausius's entropy of process starting at Q_1 and finishing in Q_2 is proportional to

$$E_c = \int_{Q_1}^{Q_2} T^{-1} dQ \tag{8.1}$$

which gives (after substitution and omission of non-essential coefficients k_Q and k_T)

$$E_J = f_J - 1 \tag{8.2}$$

and

$$E_I = 1 - f_I . \tag{8.3}$$

Considering these relations and (5.9) one obtains relation

$$E_J - E_I = f_J - f_I \geq 0 \tag{8.4}$$

which can be interpreted as an information analogue of the Second Thermodynamic Law:

1. The Ideal Gnostic Cycle (IGC) is irreversible.
2. Once the uncertainty entered into quantification, the change of entropy within the IGC is positive.
3. An information perpetuum mobile as a machine producing the information cannot exist.

The difference $f_J - f_I$ will be called the *residual entropy* of the IGC. We applied here the word "information," because it will be shown in the sequel, that there is a gnostic kind of information related to the gnostic data characteristics and their thermodynamics.

A note is in order here related to the application of entropy as the optimality criterion in algorithms. By minimization of the data uncertainty f_J one minimizes the quantifying entropy E_J and with it the whole uncertainty of the IGC. A similar effect is achievable by minimization of the difference $f_J - f_I$ or by maximization of f_I. However, the results will not be the same because of different robustness of f_J and f_I with respect to outlying and inlying disturbances.

8.2 Maxwell's Demon

To test the universality of the Second Law of Thermodynamics from the position of his "mechanical" theory of gas, James Clerk Maxwell invented a being capable to manipulate with individual molecules in an artful manner [104]. He placed this Demon into a box divided by a partition with a weightless window and charged him with duty to be a gatekeeper. Both sides of the partition were full of gas initially at one common temperature. According to the microscopic theory of temperature, there was a large spread of speed of molecules, even though the average temperatures were equal. The task of the Demon was to open the window for the fast molecules moving from the right chamber to the left one. To keep the number of molecules in both chambers equal, he let pass the slow molecules from the left to right chamber. The right chamber was becoming cooler, the left one warmer. The result was violating the Second Law: The heat was transferred from the cooler right chamber to the warmer left one.

This story could not be called a Gedanken-experiment at which all steps correspond to the Laws of Nature and are—at least thoughtfully—realizable. However, the lower power of the Second Law under the First Law was not demonstrated, because such an exception from the Law of Energy-momentum Conservation Law does not exist. But a great idea useful for this book was that of possibility of conversion of entropy to information: the Demon was getting the information on the speed and direction of molecules by his observation and using it for a decision how to manipulate the window. Let us investigate this idea in more detail and in a more realistic way.

8.3 Entropy ↔ Information Conversion

By choosing the proper values of the true value and of the coordinates $[\cosh(2S\Phi), \sinh(2S\Phi)]$ or $[\cos(2S\phi), \sin(2S\phi)]$ one can attach a gnostic event to each point of the Minkowskian or complex plane. Doing this, one attaches to the point the thermodynamic state of the event characterized by entropy. Both planes can be thus considered as scalar entropy fields of variables E_I and E_J. At invariant values of the true value, they both will be functions of the variables $S\Phi$ and $S\phi$. The field's source of the field α can be calculated by the Laplace's operator ∇_c^2 which for twice differentiable function q of coordinates x and y has the form of

$$\nabla_C^2(\alpha) = \frac{\partial q^2(\alpha)}{\partial x^2} + C^2 \frac{\partial q^2(\alpha)}{\partial y^2} . \qquad (8.5)$$

Denote

$$r_C^2 = x^2 - C^2 y^2 \qquad (8.6)$$

the squared radius of a circular path corresponding to the case denoted by indeterminate C and

$$r_{(C)}^2 = x^2 + C^2 y^2 \qquad (8.7)$$

its complement (the other case). Using the quantities h_J (2.10) and h_I (2.16) introduce two variables

$$p \equiv p_I = (1 - Jh_I)/2 , \qquad (8.8)$$

$$p_J = (1 - Ih_J)/2 \qquad (8.9)$$

where p is a real number from the interval $[0, 1]$ and p_J is a double (complex) number. Apply (8.5) to entropy E_C and multiply it by r_C^2 to make the expression valid for circles of unitary radius to rid off it from their radius equal to the true value. Two equations are thus obtained:

$$r_I^2 \nabla^2(E_J) = \frac{1}{p(1-p)} \qquad (8.10)$$

and

$$r_J^2 \nabla^2(E_I) = \frac{1}{p_I(1-p_I)} . \qquad (8.11)$$

Equations (8.10) and (8.11) are worth for further analysis. There is a source of entropy field on their left side. What if the right hand side expression would be a source of another field the changes of which would compensate the entropy's changes? To test that, integrate it twice by the variable p (p_J) to get the functions

$$I_J = -p \log(p) - (1-p) \log(1-p) \qquad (8.12)$$

and
$$I_I = -p_I \log(p_I) - (1 - p_I)\log(1 - p_I) \ . \tag{8.13}$$

The function I_J is a real function of the real argument p, but I_I is a complex function. These functions will be called *the information and the anti-information of a single item of uncertain data*.

8.4 Albert Perez's Information

To be sure in acceptance the quantity I_J for information it is useful to test it by using Albert Perez's[1] characteristics of information. As items AP will be Perez's formulations from [92, 93] and MG the corresponding statement of mathematical gnostics:

AP1 The aim of information theory is to investigate the notion of a message abstracting from the physical form of its bearer ("signal") and to reduce the disturbing impacts ("noise") which depreciate the message.

MG1 The message is the value of a true data item. The noise depreciating the observation is represented by the uncertainty manifested by inner disturbances or improper data values.

AP2 The notion of a message assumes the interaction of two systems of the observed one and of an observer or its instrument.

MG2 The system to be observed is a pair of groups of material quantities and the observer is a pair of groups of numbers.

AP3 This interaction is implemented by an exchange of energy and by a change of the system's state.

MG3 The interaction is implemented by a movement along the path's as close as possible to that of IGC.

AP4 States changes are describable by the balance of energy and of entropy with its dual, information.

MG4 States changes are described by the balance of energy and momentum of relativistic particles attached to data uncertainty by the isomorphism $\mathcal{ISMB}$ and by entropies E_I (8.9) and E_J (8.11).

AP5 The basic aspect of a message is the fidelity of reflection of the state of the source of the message by the receiving system.

---

[1] Albert Perez was a Greek scientist who became a top expert of information theory of the Institute of Information Theory and Automation of the Czechoslovak Academy of Sciences, Prague.

MG5 There are four gnostic measures of fidelity of the message, quantifying and estimating data uncertainties and certainties, relevance and irrelevance.

AP6 Entropy generally increases during the observation process creating the information, but this information can be used in partial compensation of the rise of information.

MG6 The entropy ↔ information conversion and its reverse process are described by Equations (8.9) and (8.11) as the balance of sources of fields of entropy and information.

AP7 The randomness of the impacts of disturbing effects on the message makes the description of the systems interaction by a fixed transformation impossible and imposes the application of conditional probability.

MG7 The probability entered the analysis as the quantity p in (8.12), which appeared on the places of probability in the Boltzmann's and Shannon's entropy.

AP8 The function used for evaluation of the amount of information must satisfy at least the following conditions:

1. non-negativity,
2. increasing with easiness of recognition of the state of the source of message and with refinement of the state of disturbing components,
3. reaching the minimum at equality of the probabilities of all states and being zero just at independence of the received message on the message originally sent,
4. impossibility of its increase by a measurable transformation,
5. additivity.

MG8 Formula (8.12) is a special case of Boltzmann's and Shannon's formulae for only one probability and its complement and as such satisfies all the conditions.

AP9 As a substantial element of cybernetics the information must be usable on practice to reaching goals of control with maximum suppression of disturbing impacts and to creation of machines implementing the control process on an always improving level of auto-adaptation to the features of messages and to the environmental impacts.

MG9 The approach of mathematical gnostics was tested in applications to tasks of many application fields during the decades of its existence. The results confirmed its priority over other methods of information treatment.

Statistical Interpretation of Gnostic Data Characteristics 67

A substantial comment is to be added to this comparison of Perez's requirements to information and the way how mathematical gnostics satisfies these conditions. The Perez's information was related to messages born by a multitude of data while the case considered by mathematical gnostics was dealing with a unique data item.

Another comment is necessary with respect to the anti-information I_I. It is a complex function of the complex double number p_I. Its value is minimum with zero parameter p_I and unlimitedly increasing with rise of its module. It might be interesting to physicians studying the quantum mechanics.

On the other hand, the anti-information may be also useful for estimation thanks to formula (5.9) bounding together quantifying characteristics with the estimation one. The hyperbolic tangent is zero along with zero anti-information and tends to the limit value when anti-information goes to infinity. The value of anti-information is thus measurable by the same probability as the information. Gnostic probability is thus unique.

The additivity has been accepted for both data uncertainties and irrelevances. The probability is a linear function of the estimating relevance. Additivity of gnostic probability is thus warranted as well.

8.5 Statistical Interpretation of Gnostic Data Characteristics

The convergence of both uncertainty f_J and certainty f_I to 1 with Z approaching Z_0 as well as the convergence of both irrelevance h_J and relevance h_I to zero in the same case have certain important consequences. Let ξ be a real number, N an integer and $O(\xi^N)$ be the so-called Landau's symbol characterizing the order of magnitude. This symbol is equivalent to the statement, that for ξ converging to zero, the variable $O(\xi^N)$ converges to zero as ξ^N. Expanding the formula $\phi(S\Phi) = \arctan(tanh(S\Phi))$ into the Taylor series, one can easily verify the relation

$$S\phi = S\Phi + O((S\Phi)^3). \tag{8.14}$$

Taking the same approach to 2.9 and 2.15 one obtains

$$f_J = 1 + \frac{(2S\Phi)^2}{2} + O((S\Phi)^4) \text{ and } h_J = 2S\Phi + O((S\Phi)^3). \tag{8.15}$$

$$f_I = 1 - \frac{(2S\Phi)^2}{2} + O((S\Phi)^4) \text{ and } h_I = 2S\Phi + O((S\Phi)^3). \tag{8.16}$$

Let us use the term *a sufficiently precise datum* to imply a datum, for which the (additive) error $S\Phi$ permits the terms $O((S\Phi)^3)$ and $O((S\Phi)^4)$ to be neglected. Then these relations prove, that the following statements are valid *for sufficiently precise data*:

1. The difference between the Euclidean and Minkowskian angles $S\phi$ and $S\Phi$, which evaluate the data errors, tends to zero when the statistical error becomes sufficiently small.
2. Both gnostic evaluations (irrelevance and relevance) of the data error tend to the additive error $2S\Phi$.
3. Both gnostic data certainty and uncertainty tend to a simple quadratic functions of the additive error.

The nature of the quartet of gnostic data characteristics is thus reflected by their names: the certainty and uncertainty evaluate the data deviations from the "ideal" (true) value like data variance in statistics. The relevance and irrelevance say, how relevant are the observed values for the judgment on the true value like linear data errors do in statistics of sufficiently precise data.

It is obvious from the above general relations, that all these statements hold, if and only if data are sufficiently precise.

There are important consequences of these relations for other gnostic evaluations of uncertainty, entropy, probability and information. To show the behavior of gnostic characteristics under conditions of uncertainties of different size, it is useful to consider the relative estimating error introduced already in (4.1):

$$S\Phi_k = (A_0 - A_k) \qquad (8.17)$$

where A_0 is again the true and A_k the observed value of the k-th data item, and where S is the same scale parameter as before. Four classes of data errors by size are defined in Tab. 8.1.

TABLE 8.1
Four classes of data errors

Error's size	Symbol of the class	Approx. bounds
Very small errors	VS	$|S\Phi_k| \leq 0.005$
Small errors	SE	$0.005 \leq |S\Phi_k| \leq 0.015$
General case	GC	$|S\Phi_k| < \infty$
Extreme	EX	$|S\Phi_k| \to \infty$

The numerical values of bounds of errors are in a way arbitrary, the smaller are they chosen, the more exact the approximation is shown in the next table.

This small table is worth for a careful examination, as it summarizes the results of the foregoing sections with respect to estimation process. The fourth column (the general case **GC**) identifies each variable with the applicable general formula, which is valid for an error of an arbitrary size. The other columns contain approximations obtained for the special cases of data size, which were defined in Tab. 8.1.

Very small data errors (VS): *The estimating error* evaluated by the irrelevance approaches the value of the relative error, which is one of the traditional evaluations of the error of multiplicative data. (The constant scale factor 2 does not play a significant role.) *Data relevance* is a constant, independent of the error value. All data have the full relevance of 1. *Probability* equals 0.5: all we can get from the datum's value is, that the unknown ideal value may be either less or more than the observed data value. This conclusion does not depend on the data error. *Entropy changes* as well as *information changes* are completely ignored.

Small errors (SE): *Error evaluation* is the same as for very small errors, but the *data relevance* decreases with increasing squared error—worse data are getting a smaller relevance than the better ones. The *probability* of the true value is a linear function of the relative error. This permits a rough characterization of the probability's dependence on the error of data values close to the ideal value to be obtained. The *entropy* and *information changes* are evaluated by the same formula, but with the opposite signs. Entropy changes are thus completely balanced by the information changes—the quantification/estimation cycle is (approximately) **reversible** in this special case. The quadratic character of both functions of uncertainty supports the application of the least squares method for data contaminated with small errors.

Extreme errors (EX): The bounds for errors obtainable for gross data errors and outliers are obtained as limits of the general case (GC) and are shown in the last column of Tab. 8.2. The most important fact is, that all estimating characteristics are **bounded**—an unlimitedly increasing or decreasing data error (outlier) cannot force an estimating characteristic beyond its finite range. This feature has already been identified to be the source of robustness with respect to outliers, which characterizes a class of gnostic estimating procedures.

Analogous table could be shown relating to the quantification process.

TABLE 8.2
Estimation characteristics of data uncertainty for different classes of data errors

Estimation characteristics	Class of the error			
	VS	SE	GC	EX
Data relevance	$2 \cdot (S\phi_k)$	$2 \cdot (S\phi_k)$	$h_{I,k}$ (2.10)	-1_+ or $+1_-$
Data error	1	$1 - 2 \cdot (S\phi_k)^2$	$f_{e,k}$ (8.3)	0_+
Entropy fall	0	$-2 \cdot (S\phi_k)^2$	$f_{e,k} - 1$ (8.3)	-1_+
Probability	$1/2$	$1/2 - (S\phi_k)$	p_e (8.8)	0_+ or 1_-
E-information	0	$2 \cdot (S\phi_k)^2$	$I_{q,k}$ (8.12)	$\log(2)$

The additive error $S\Phi$ is a linear function of the (unknown) ideal value A_0 (4.1). The estimate of this quantity can be obtained easily by minimizing the sum of squares of additive data errors. This is the well-known *ordinary least squares* (OLS) statistical estimating methodology in which an estimate is *unbiased* (the sum of estimating errors equals zero). It is obvious from the foregoing statements, that the gnostic characteristics, which have been considered, approach this most popular and frequently used statistical technique (additive errors and their squares), if the data are sufficiently precise.

Conclusions are clear: gnostics is not "something against statistics." It is an autonomous approach suitable even for small data samples. On the contrary, it justifies the application of statistics to data that are sufficiently precise (as shown by the Tab. 8.2) by means of mathematical reasoning entirely different from the statistical ones. However, as is also demonstrated by this table, handling of strong uncertain data is to be entrusted to algorithms differing from those of classical statistics.

According to the Kuhn's "theory of theories" a more general new paradigm should give the same results as the old one when applied to some special cases. It results from the foregoing analysis, that mathematical gnostics aspires to be acknowledged as a generalization of statistics valid even for strong uncertainties of small data samples.

8.6 Between Mediocristan and Extremistan

This section is a citation of a paragraph of unpublished book [40].

Two provinces of real world are distinguished by N. N. Taleb in his famous book Black Swan [37] defined there in the following way:

Mediocristan: the province dominated by the mediocre, with few extreme successes or failures. No single observation can meaningfully affect the aggregate. The bell curve is grounded in Mediocristan. There is a qualitative difference between Gaussians and scalable laws, much like gas and water.

Extremistan: the province where the total can be conceivably impacted by a single observation.

The main distinction between the provinces is here the impact of a single observation on the total (aggregate). From the point of view of data analysis, it can be useful to modify the picture risking to be considered as "the Platonified scholar who needs theories to fool himself with" (N. N. Taleb's words).

Considering uncertain data, one should take in account quantitative factors especially the amount of event's uncertainty and number of events. The role of uncertainty is characterized in Tab. 8.2 by distinguishing between four classes of data errors and demonstrating the convergence of gnostic parameters of uncertainty to classical statistical ones. Strong uncertainty makes the space of uncertain data curved and the impact of quantifying uncertainty unlimitedly large. The "total" of small number of events is more sensitive than the larger one. Conditions of weak uncertainty and "sufficiently" large number of events make classical statistics applicable and enable accepting of membership in Mediocristan, but their absence excludes them from this category. A special problem is that of predictability. Two necessary resignations are necessary before a would-be forecasting: a hope to predict a Black Swan and to predict an uncertain event precisely. On the other hand, it would be a pure agnosticism to deny the predictability completely. Even imprecise prediction can be profitable as demonstrated by prosperity of the firms engaged in algorithmic programming.

This is why the notion of "statistical error" has been introduced. All four error classes shown in Tab. 8.2 allow the imprecise prediction and for the category "Extreme" does not include the Black Swans, because the table relates only to the uncertain events considered within the framework of gnostic theory. This table inspires an alternative sub-division of the "uncertain data world":

Mediocristan: A non-curved space of idealized uncertain data.[2]

Ordinarystan: A curved space of real uncertain but probable data.

Extremistan: Data on highly improbable events such that one single observation can disproportionately impact the aggregate or the total.

The geometry in Mediocristan is Euclidean, independent of the data location in the space. Typical events are mechanical events viewed by Newtonian/Gallileian eyes or statistical events assumed by the Central Limit Theorem.

Observability of real data is determined in gnostic theory by the first axiom which requires of both true and disturbing data components being members of given algebraic structures. Data from Ordinarystan include not negligible uncertain components which make the space curved and metric location-dependent of Riemannian type. Typical members are data sets investigated in robust statistics and in gnostic theory. The definition of Extremistan is Taleb's. Such data are not considered in gnostic theory. Events producing data which belong to both Mediocristan and Ordinarystan are "imprecisely" (more or less successfully) predictable.

---

[2]In case of a bi-dimensional space, such a space could be called Flatland remembering E. A. Abbott [5].

8.7 Conclusions

The thermodynamical interpretation of data certainty and uncertainty, relevance and irrelevances enables the quantifying and estimating change of entropy as two measures of uncertainty to be evaluated. A statement can be thus formulated analogous with respect to the Second Thermodynamical Law: The Ideal Gnostic Cycle is irreversible and the overall change of entropy within it is positive. The idea of famous Maxwell's virtual demon shows the possible interdependence of entropy and information. The analysis of the two fields of gnostic entropy enables the derivation of equation of sources of entropy fields balanced by sources of another fields which appears to be a real field of information and a complex anti-information field. A series of nine requirements to notion of information formulated by A. Perez is satisfied proving the justification of application of the word information to the case of a single item of data considered by mathematical gnostics. Two unique notions of gnostic probability are derived as the information's and anti-information's parameter.

When the considered data are sufficiently precise, the gnostic characteristics of data uncertainty tend to statistical ones. It means, that mathematical gnostics can be considered as an extension of statistics valid for small data samples and strong uncertainties.

9
Kernel Estimation

9.1 Parzen's Estimating Kernel

Parzen's method of kernel estimation [91] of probability density has old roots in mathematics (e.g. convolution of Green's functions or Duhamel's convolution integral). Unlike his predecessors, who could apply known special solutions of their equations to play the roles of "kernels," no statistical kernel was available to Parzen. He therefore proved conditions for a kernel to produce a density estimate converging to the true one. Parzen's kernel $K(x)$ should satisfy the following conditions:

$$\sup_{-\infty < x < \infty} |K(x)| = \infty \;, \tag{9.1}$$

$$\int_{-\infty}^{\infty} |K(x)| < \infty \;, \tag{9.2}$$

$$\lim_{x \to \infty} |xK(x)| = 0 \;, \tag{9.3}$$

$$\int_{-\infty}^{\infty} K(x)dx = 1 \;, \tag{9.4}$$

$$(\forall x \in R^1)\; (K(-x) = K(x)) \;. \tag{9.5}$$

The *kernel estimate* of the probability density is thus obtained as

$$g(x) = \frac{1}{NS} \sum_{1}^{N} K\left(\frac{x - X_n}{S}\right) \tag{9.6}$$

where X is a data sample and S its scale parameter. Parzen's kernel estimation is recently available in many packages of statistical programs, but with problems:

- There are many forms of kernels satisfying Parzen's conditions producing estimates of different (even unsatisfactory) quality when applied to real (finite) data samples. A narrower definition of a kernel is desirable.

- Some of Parzen's kernels are not differentiable.

- Additive aggregation of kernels applied in Parzen's approach misses a theoretical justification and is not always the best one.

DOI: 10.1201/9780429441196-9

- A constant, arbitrarily chosen form of kernels (especially their width determined by S), can be inadequate for some data samples.
- Application of a constant width of all kernels of a data sample can be in a contradiction with the heteroscedasticity of some data.

There also are significant positive aspects of the kernel estimation:
- An a priori guess on the "fitted curves" (or on "standard types" of a probability distribution to be estimated) is not required.
- Probability densities of forms far from some simple analytically defined curves or standard statistical distributions can be estimated.

In other words: kernel estimates can be closer than others to the ideal saying "Let data speak for themselves."

9.2 Gnostic Kernel

Return to the function p (8.8) using the substitution (2.16) to obtain
$$p = (1 - \sin(2S\phi))/2 \tag{9.7}$$
which shows the variable $p(\phi)$ as a probability reaching zero for $S\sin(2\phi) = 1$ and 1 at $\sin(2S\phi) = -1$. Its derivative is
$$\frac{dp(A)}{dA} = \frac{1}{S\cosh^2(2S\Phi)} \tag{9.8}$$
which satisfies all Parzen's conditions.

Mathematical gnostics comes with several innovations improving the kernel estimation:

1. A unique, smooth and differentiable kernel (9.8) satisfying Parzen's conditions has been obtained by analysis of the quantity p which appeared as the argument of information. This function of A will be called *gnostic kernel*.
2. Application of this kernel leads to maximization of information because the kernel's form corresponds to the Ideal Gnostic Cycle.
3. The kernel's form and "width" is not subjectively chosen but theoretically justified and practically estimated from data respecting the curvature of the space of uncertain data.
4. The data heteroscedasticity of data can be taken into account by giving different (theoretically justified) scale parameters to different data values.
5. There are two theoretically justified ways of kernels aggregation to get distribution functions: the mean value of the kernels' estimating certainties and the mean value of the kernels' quantifying

uncertainties ensuring two types of estimate's robustness (with respect to outliers and/or to inliers and noises) to be obtained. These *types* of distribution functions will be called *estimating distribution*, denoted *EGDF* and *quantifying distribution*, denoted *QGDF*.

6. There also are two *kinds* of distribution functions to be obtained: the *Local* ones denoted *ELDF* and *QLDF* which use the simple additive aggregation (9.6) of kernels and the *Global* ones denoted *EGDF* and *QGDF* which are obtained by *normalized aggregation*.

The notion of normalized aggregation deserves a comment. The additive aggregation (9.6) provides the mean value of kernels. Each kernel is a probability density and their integral is the probability. However, mean kernel is not a probability density because mean sinus is not a sinus. Nevertheless, it is useful in revealing interesting properties of data structure. But to make the mean kernels the probability density requires its normalization by a suitable divisor.

9.3 Scale Parameters

The scale parameter S plays several important roles in data treatment:

- It determines the width of the gnostic kernel both in (9.6) and (9.7).
- It controls the width of characteristics of all individual data items deciding thus on degree of smoothness of results of aggregation.
- Its variability decides on homo- or heteroscedascity of a data sample.
- It enables to judge on the curvature of the space of uncertain data.

It is therefore important to be able to estimate scale parameters. A theorem from [62] says the following: Given a data sample of homogeneous data Z transformed to the standard data domain a sub-sample of which has a constant scale parameter S. Given a fixed point Z_0. It is required to find such S that the following equation holds:

$$\int e_J(z/Z_0)\,dp(z/Z_0) = \overline{(e_J(Z_k/Z_0))} \qquad (9.9)$$

where the overline denotes the arithmetical mean of the expression and when index k goes over the data of the sub-sample. The equation requires the equality of the mean entropy e_J calculated by its continuous model with the point estimate using sub-sample's data. The theorem says that a unique solution exists coinciding with the solution of equation

$$\frac{\pi S Z_0/2}{\sin(\pi S Z_0/2)} = \overline{(f_J)} \ . \qquad (9.10)$$

An analogous theorem can be proved for the quantification entropy e_I and the uncertainty f_I. Unlike the estimation case where for the S exists a maximum value of 2, the quantifying S can rise to infinity. This substantial difference makes the Equation (9.10) useful but with a care related to the scale parameters.

The forms of the gnostic kernel is viewed in Fig. 9.1. The scale parameter S determines the width of the kernel, the less S is, the narrower the kernel.

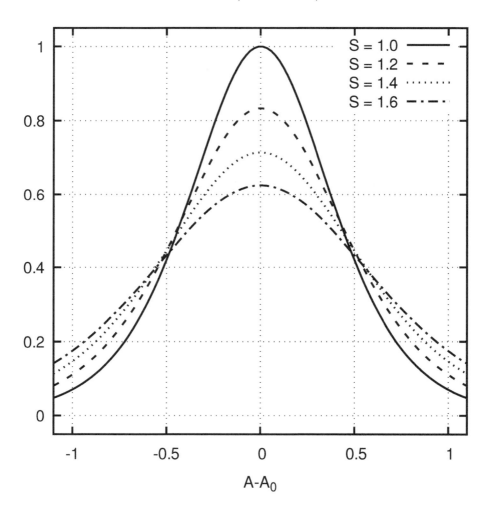

FIGURE 9.1
The gnostic kernels in dependence on scale parameter S

9.4 Conclusions

The statistical methodology of kernel estimation developed by E. Parzen provided chances to estimate the probability and its density of data samples not having a sure distribution function of a standard (named) type. But Parzen's conditions for a suitable form appeared to be as weak as allowing a large amount of candidates for the kernel to be designed and applied. The unfavorable experience with many kernels represented a call for a theory which would be able to uniquely choose and justify the suitable form of the kernel and its aggregation suitable for necessary sorts of applications. The suitable form of the kernel was found in gnostic probability density function of a single item of uncertain data.

The existence of two types of gnostic data certainties and uncertainties and relevances and irrelevances offering two types of estimating robustness opened the door to two types of estimates of distributions of probability and its density. However, an additional requirement of having the means of kernels representing the uncertainty of data characteristics led to introduction of two kinds of distributions, the local and global ones distinguished by the additive and normalized aggregation of kernels. Four estimates of gnostic probability distribution functions are thus to be discussed.

The gnostic distribution functions of probability and density enable rich applications of the gnostic kernel to be implemented. Moreover, the gnostic kernels appear to be a mighty instrument in investigation of general relations between uncertain variables.

The scale parameters estimated by data values play an important role in analysis.

10
Probability Distribution Functions

10.1 Probabilities

The real variable p_I (8.8) has a value from the interval $[0,1]$ because the estimating irrelevance h_I is from the interval $[-1,1]$. Moreover, it appeared in the formula of information I_J (8.12) in the role played in Boltzmann's and Shannon's formulae by probability. The relevance is a feature of one data item and is additive. It means that the "collective" relevance of a data sample can be expressed as a mean value of individual relevances. But remember the mapping $\mathcal{ISMA}$ which maps data with relativistic particles. The "collective" relevance will be mapped as the mean of momenta of particles and the mean certainty as the mean energy of particles. It is not the same like the idea of only one particle representing the collective by its momentum and energy (one single data item representing the sample by its relevance and certainty) because mean of sines is not a sine and mean of cosines is not a cosine. To make the mean relevance and mean certainty of L data item representing a single data item as f_I and h_I, the equations

$$f_I = \sum_{i=1}^{i=L} f_{i,I}/N_I \tag{10.1}$$

and

$$h_I = \sum_{i=1}^{i=L} h_{i,I}/N_I \tag{10.2}$$

where

$$N_I = \sqrt{\sum_{i=1}^{i=L} f_{i,I}^2 + \sum_{i=1}^{i=L} h_{i,I}^2} \tag{10.3}$$

must hold with N_I as a "norm." To get the probability p of a data sample of L data, one has two possibilities: probability of the mean irrelevance (the *local probability*)

$$ELDF = \frac{1 - \sum_{i=1}^{i=L} h_{i,I}}{2L} \tag{10.4}$$

and mean probability of the sample (the *global probability*)

$$EGDF = \frac{1 - \sum_{i=1}^{i=L} h_{i,I}}{N_I}. \tag{10.5}$$

Each relevance and certainty is a function of three variables: additive A and A_0 or multiplicative Z and Z_0 and of the scale parameter S. The gnostic kernel's value (9.8) can be considered as a value of a distribution function given A, A_0 (or Z and Z_0) and S. When fixing the true value A_0 and considering different values of the A, we obtain the probability density distribution of the kernel. Considering the Equations (10.4) and (10.5) we see, that they represent probability distributions obtained by integrals of gnostic kernel given the true value. We shall call the *ELDF* the *estimating local distribution function* and the *EGDF estimating global distribution function*. A question might result considering the both types of distributions obtained by integrals of gnostic kernels and not directly by kernels. The answer is that applications frequently include cases of data with an inconstant scale parameter, the *heteroscedastic data*. To evaluate the probability density function of such data analytically would include complicated derivative operations. Instead, to make solution of such problems available, the density function is obtained by numeric derivative of the probability obtained not by density kernels like Parzen did but by their integrals.

To estimate the quantifying probability distributions (local $QLDF$ and global $QGDF$) the relation (5.6) is used determining the estimating relevance by means of the quantifying and not estimating functions. The formulae are as follows:

$$QLDF = \frac{1 - \sum_{i=1}^{i=L} H_{i,J}}{2L} \tag{10.6}$$

and

$$QGDF = \frac{1 - \sum_{i=1}^{i=L} H_{i,J}}{N_J} \tag{10.7}$$

where

$$N_J = \sqrt{\sum_{i=1}^{i=L} 1/(f_{i,J}^2) + \sum_{i=1}^{i=L} H_{i,J}^2} \tag{10.8}$$

where

$$H_{i,J} = h_{i,J}/f_{i,J} \tag{10.9}$$

by (2.14).

10.2 Data Domains

The original data model (4.1) was additive. Such data can be called *additive*, as they are members of the additive version of the Abelian group. The

Data Domains

exponential operation as (4.2) makes the data multiplicative. The former data are real numbers, the latter one are positive real numbers. The observed data may be of both kinds but to treat them we require a standardization. We will therefore distinguish the original data domain calling it *natural domain* and indexing it by a (*additive*) and m (*multiplicative*). As the finite data domain used for the majority of data treatment operations as a *standard domain*, the closed interval $[exp(-1), exp(1)]$ will be used indexed by z. The bounds will be reached when the natural data value will equal to the least or largest value of the data sample. To represent the estimated probability distribution or its density, the distribution over the natural domain will be shown to inform the user on the probability of individual quantiles and on the form of distributions.

There will also be the infinite data domain. This point deserves a comment. The point of mathematical gnostics is that real quantities are finite. The infinite bounds of statistical models of distribution functions should be understood as an "approximation" of very large quantities. Even the fastest motion, the speed of light, has its limit. The size of cosmos is very, very large but to consider it infinite would rise a question on scientific proofs. All this means, that for data treatment one could limit oneself by considering the finite domains of things. The information obtained by estimating the bounds of data domain is frequently the best goal of data analysis. But there are good reasons to introduce the infinite data domain, because it can reliably decide on homogeneity of data samples and to get rid out of the sometimes strange effects at the edges of finite distributions. There is also a good chance in estimating the distribution over the infinite domain consisted of introducing two variables called LB and UB, (the *lower and upper bound*) which can be estimated as the bounds of the probable values of the observed variable.

A series of transformations is necessary for working with the distributions by changes in both domains and ranges:

- additive to standard domain,

- multiplicative to standard domain,

- standard to infinite domain,

- infinite to standard domain,

- standard to additive domain,

- standard to multiplicative domain.

The transformations use the least and largest sample's value LD and UD and the bounds LB and UB estimated within the process of fitting the global distribution function to the sample's data. The standard domain is the interval $[\exp(-1), \exp(1)]$.

10.3 Tasks Solvable by Distribution Functions

The following are some of the few tasks for which the distribution functions are originally created:

- to estimate probability of a given quantile,

- to estimate quantile of a probability,

- to provide the form of the distribution function for a class of events.

The connections between the distribution's form and a class of events in statistics were established by experiments and by a theory. When a particular data sample appeared the problem arose to which class of events it belongs to get the corresponding distribution and especially if such a suitable—and known—class exists. The application of kernel estimates enable to leave the idea of known classes of events and to estimate the distribution function to each individual data sample by using its data. The resulting distribution function reflects the individuality of the data sample and not of some artificial data collections. A physician is interesting more on distributions of a parameter within its own patients than on distribution of the whole population although the comparison of the former and latter can be also useful. An analyst of an enterprise is more interested in distribution of economic and technology parameters of his facility and of the competitors than in distribution—if they are available—of the whole industry. A researcher needs primarily the distributions of his own data. Data should be specifically reflecting the real state of the data source. Using the kernel estimates of distributions sets the analyst free of uncertainty if his data belongs to a class of events with a known distribution function and gives him the required results. But there are also applications of this technology solving some new tasks.

The first fundamentally important task is the decision of which distribution functions will be used. If there are doubts on applicability of a standard statistical distribution functions then the suitable kernel estimate is the clear choice.

The second unusual task of kernel estimation is that of estimating the bounds of the data support connected with data fitting procedure of design of global distribution function. Three quantities are to be found (bounds LB and UB and a mean value S of scale parameters) minimizing the criterion function.

The third purpose of using distribution function is the robust estimation of the location parameter which can be used as the sample's mean value.

The fourth category of tasks is related to the data structure. To make the data specific, the analyst establishes qualitative and quantitative conditions for the choice of objects to be measured. Only the data values can then say, if his selection was successful in providing him with data on the objects he specifically wanted. The ideal case is when the data sample brings the *homogeneous data*. The problem of homogeneity is not adequately scientifically solved in statistics although it is a real problem of sampling. The first duty of a statistician in sampling should be taking care on comparativeness of all participants of the sample to prevent comparison "of apples with pears." When the data homogeneity is proved, the real chance exists that the data sample bears information on a properly delimited choice of uncertain events. But when the homogeneity test fails, than further eliminating phase called *homogenization* must follow to get the homogeneous sub-sample or *cluster analysis* has to be done to isolate the individual homogeneous sub-samples. The non-homogeneity of a sample is manifested by multiple local maxima of the global probability density function. They may show that not one group of uncertain events has been observed but more of them. Another disturbance of homogeneity can be caused by outliers. The technology of kernel estimation has software tools for homogeneity tests as well as for homogenization and cluster analysis. The impact of outliers is minimized by the robustness of the method.

The fifth category of the new tasks for the kernel estimations is connected with existence of not completely determined data called the *censored data*. There are three kinds of such data:

- The left-censored data ($D \in (LC, \infty)$)
- The right-censored data ($D \in (-\infty, RC)$)
- The interval data ($D \in (LC, RC)$)

where LC and RC are constant. The first case are especially data measured below a sensitivity bound LC of the method. The second case includes especially data measured over the measurability range and the survival data where RC is the last moment at which the object was registered as living. The interval data are produced by some measuring devices registering the values on an discrete scale of pairs (LCi, RCi) of a series of *is* like for e. g. the multichannel analyzer of gamma rays. To include censored data, the user provides the estimating algorithms both with data D and bounds LC and RC. The motivation of using the censored data in analysis stems from the fact that such data are met without chances to get their complete form and that the applications include very important task as the solution of survival problems or concentrations of dangerous pollutants. But there will also be other useful applications of censored data demonstrated.

The sixth category is the survival analysis. The finite bounds of data domain along with the treatment of censored data enable the life-time estimation to be performed.

The seventh category is the robust hypotheses testing. Availability of kernel estimates of distribution function of an arbitrary data sample enables all reliable information necessary for the decision making by testing of hypotheses to be obtained.

There also is the eighth category of the new tasks of the kernel estimation called *interval analysis*. It uses the unexpected feature of the *ELDF* of a homogeneous data sample consisting in its complex reactions to large changes of quantile variable. This enables several natural bounds of intervals to be estimated. This technique will be described in the sequel.

10.4 The Estimating Local Distribution

The estimating local distribution function *ELDF* is obtained in a straightforward manner as an arithmetical mean of gnostic kernels provided with their individual or joint scale parameters. Based on the dependence of the scale parameters' value the form of the individual kernels is more or less smoothed. This flexibility is advantageous making it possible to reveal all details of sample's data structure, to isolate the outliers and/or individual sub-samples making a common maximum of the density function. The example is in Tab. 10.1 showing the *ELDF* of the fourth column of the data file *stackloss*[1] It is useful to show here this interesting function (Tab. 10.1).

The example shows four graphs in the form of (row, column) where the graph (2,1) shows the probability distribution over the natural (finite) domain with the density distribution in (2,2). The distributions over the infinite domain are in (1,1) and (1,2). The quadrates show values of the halves of steps of empirical step function and small circles depict their projection on the smooth distribution line. The scale parameter was set at the value 1. Five local clusters appear with this scale parameter. The number of clusters depends on the S. So, e.g. application of the $S = 0.1$ leads to showing an individual sharp cluster for each data value differing from others as shown in Figure 10.2.

The play with the scale parameter may reveal interesting clusters motivating the effort of finding the causes of clustering. This can be achieved by finding additional data or by making use of data of other columns of the matrix. Interesting results may be obtained by *marginal analysis* consisting in decomposition of the sample by isolating the data belonging to individual clusters. Forces cumulating the data into clusters frequently stem from real micro-processes within the main process producing data and their identification may be an additional gain of the analysis.

This unlimited flexibility of the estimating local distribution is unique and is not shared with other distribution functions and makes the *ELDF* a suitable instrument for data analysis and for visualization of the data structure.

---

[1] The famous data file *stackloss* available in database of the R-project (www.r-project.org) has been widely used and called "the guinea pig of robust statistics."

TABLE 10.1
Data sample *stackloss*

Air Flow	Water Temp.	Acid Conc.	Stack.Loss
80	27	89	42
80	27	88	37
75	25	90	37
62	24	87	28
62	22	87	18
62	23	87	18
62	24	93	19
62	24	93	20
58	23	87	15
58	18	80	14
58	18	89	14
58	17	88	13
58	18	82	11
58	19	93	12
50	18	89	8
50	18	86	7
50	19	72	8
50	19	79	8
50	20	80	9
56	20	82	15
70	20	91	15

10.5 Quantifying Distributions

To obtain the quantifying version of distribution function requires to find the suitable variable for the probability because the anti-probability function is a complex variable with an infinite range. Such a suitable variable is offered by formula (5.6) which enables to view the probability both as an object of estimation (defined by the argument ϕ/S) and as an quantification object (defined by the argument Φ/S). They will differ a.o. by different sensitivity to the "bad" data (by robustness.)

The *QLDF* of the same data file as above for $S = 1$ is shown in the 10.3. The *QLDF* format is much more rigid than *ELDF*, therefore the distribution is more smooth. It is not suitable for the visualization of the data structure.

The methods of data treatment are dependent on features of data, on their additivity or multiplicativity, on the type of suitable metric (quantification (Q), estimation (E) or residual entropy (R)) and on their homo- or

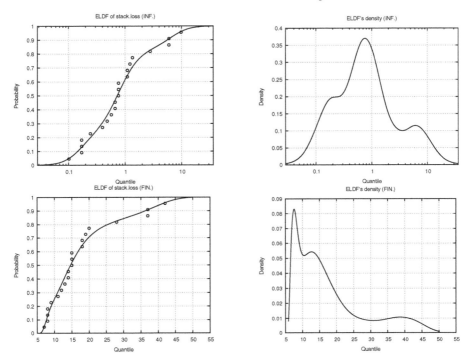

FIGURE 10.1
Distribution function $ELDF$ of the sample *stackloss* with $S = 1$

heteroscedasticity (constant or variable scale parameter). These are given as following parameters:

$Add = TRUE \mid FALSE$,

$Type = Q \mid E \mid R$

$VarS = TRUE \mid FALSE$

These parameters are not assumed a priori but are automatically decided by data analysis. Data are tested by an algorithm comparing all combinations of these parameters to find the most suitable triad of parameters.

10.6 Empirical Distribution Function and the Fit

The smooth cumulative distribution functions are obtained by fitting the well-known empirical distribution function. When applying the Kolmogorov-Smirnov statistical test of goodness of fit of a smooth model to the

Empirical Distribution Function and the Fit

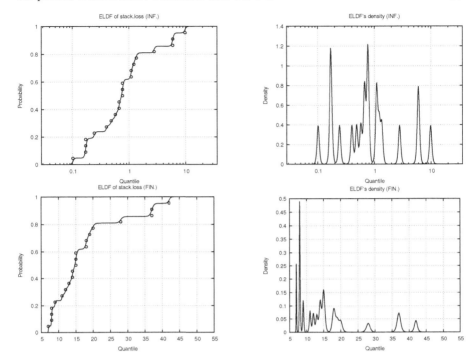

FIGURE 10.2
Distribution function $ELDF$ of the sample *stackloss* with $S = 0.1$

empirical distribution function one controls the maximum ("vertical") distance between the points on the smooth line and the point lying on the midpoint ("K-S point") of the step function. After doing this for all data, the over-all maximum is compared with the Kolmogorov-Smirnov statistics to judge on the goodness of fit. The gnostic fit is entirely different:

- Order the data sample $D_1, \ldots, D_L$.

- If the data are additive, transform them into the standard interval.

- Design the K-S points $1/(2L), 3/(2L) \ldots (2L-1)/(2L)$ and transform them onto the infinite data domain parameterized by free parameters LB and UB.

- Prepare the smooth function by aggregating the normalized mean of gnostic kernels also transformed onto the infinite data domain.

- Define the criterion function CF as the mean distance between the KS-points and the points of the CF lying in NW- or SE-direction from the KS-points (perpendicularly to the smooth line).

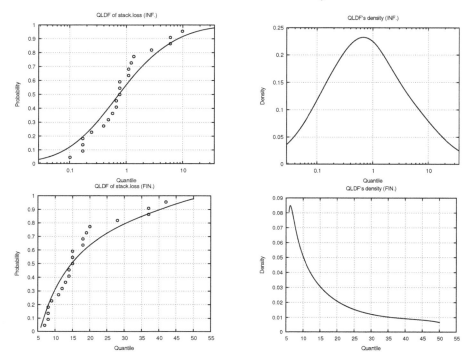

FIGURE 10.3
Distribution function $QLDF$ of the sample *stackloss* with $S = 1$

- Run the minimization procedure

$$\min_{LB, UB, S} (CF) \tag{10.10}$$

where LB and UB are the probable bounds of the probability domain and where S is the mean of scale parameters.

- Transform the cumulative distribution and its density back onto the natural data domain and print the results.

Instead of verifying the goodness of a particular fit, an optimum fit is construed. Instead of taking into account only the vertical distance (the probability error) and ignoring the horizontal deviation (the quantile error), both errors are accounted for and also minimized.

It is useful to apply some real data in examples of distribution functions. The famous function *stackloss* is chosen given as a matrix:

An example of the $EGDF$ is in Fig. 10.4 where the best fit of the sample $stackloss[,4]$ by the $EGDF$ is shown. The bounds $LB = 4.77$ and $UB = 109.7$ of the probability domain resulted with the optimum of the scale parameter $S = 1.56$. When checking the course of the density in the graph (1, 2) by

Empirical Distribution Function and the Fit 89

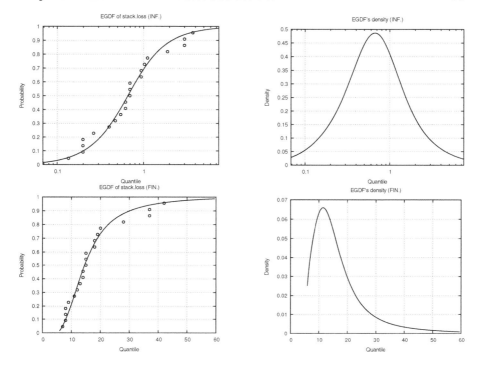

FIGURE 10.4
Distribution function $EGDF$ of sample $stackloss[,4]$ with optimum $S = 1.56$

eye, the sample has been found as homogeneous (there is only one density maximum over the infinite axis).

However, the optical control of the density curve is not sufficiently reliable, therefore the digital control is necessary. So, for e. g. when the lowest value of the sample (7) is decreasing to 2.7, the sample becomes non-homogeneous, because digital value of the density over the infinite axis stops its rise at the value $2.463 \cdot 10^{-3}$ and falls to the minimum $2.284 \cdot 10^{-3}$. Such a fine change cannot be observed on the graph the maximum of which is about 0.8.

The distribution $QGDF$ of the same sample is shown in Fig. 10.5.

To compare the forms of distribution functions optimized by using different metrics, Tab. 10.2 can be used.

TABLE 10.2
Comparison of parameters of two distributions

Distribution	S	LB	UB	LP	$MAPE$
$EGDF$	1.56	4.77	109.7	11.41	0.0303
$QGDF$	2.02	5.47	82.5	11.38	0.0315

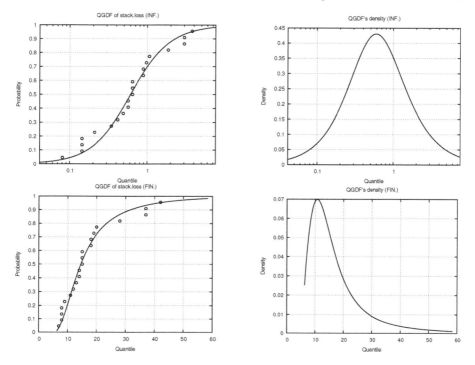

FIGURE 10.5
Distribution function $QGDF$ of the sample *stackloss*

The symbol S denotes the scale parameter, UB is the lower and UB upper bound of the data support, LP is the location parameter (quantile of the maximum density) and $MAPE$ means Mean Absolute Probability Error. When judging the results of the comparison, it is necessary to take into account that the two distribution functions are robust in a different way and they suppress different parts of disturbing data. A rational approach is to prepare both types of distributions and then to use all additional information to finally choose the "better" one.

10.7 Some Applications of Distribution Functions

10.7.1 Revealing Historical Information

Collecting of historical coins is a pleasant entertainment but may also be a reliable way of getting historical information not only of iconographic value but also on the economic development of countries. An example is connected to

TABLE 10.3
Number of coins analyzed for weights and purity and ranges of the values

Mintage		Weight (g)		Purity (rel.unit)	
Type	Time	Range (g)	Pieces	Range (1)	Pieces
I.	1346–1348	3.235–3.645	14	0.843–0.875	3
II.	1348–1355	3.204–3.490	3	0.875	1
III.	1350–1358	3.125–3.520	13	0.847–0.870	7
IV.	1358–1378	2.430–3.586	13	0.838–0.865	9
V.	1370–1378	2.480–3.787	24	0.749–0.767	7

collections of Prague Grossis [102] that was a reputable money favored in whole Europe during Middle Age. Measuring of weights and purity of coins belong to basic ways used in numismatics along with the iconographic identification of coins. However, the distribution functions enable additional information to be obtained, especially due to their applicability to small data samples. The coins used in cited study are identified in Tab.10.3. The following text is a citation from the paper [102].

Grossi Pragenses

The name for these coins is, of course, Latin and it is derived from Prague Grosh[2]. Its use became popular in Europe because of the economic power of the Czech kingdom and its adoption as national currency by several countries. The coin, minted in Czech silver was embossed with the Latin name of the Czech king, Vienceslaus Secundus (Václav II.), on one side and with the Czech crown and double tailed lion on the other. The coin was introduced in 1300 during the economic reforms instigated by King Václav II (1278–1305) and the coinage was called "eternal and holy" probably because of its stability and universal liquidity stemming from its broad use in Europe. It is estimated, that annual production of silver from only the Kutná Hora[3] mines was around 20,000 kilograms. The coin maintained its value primarily due to its silver content and mint marks, which guaranteed its authenticity. This contributed to its continued employment in international trade. Due to its silver content, it was in circulation for several centuries until the quality of the coins deteriorated. Collections of these coins, which exist today, mostly stem from archeological sources, and unfortunately, only a few of them have been quantitatively described so as to conduct a thorough analysis. It is something of a paradox, that numismatists are ready to spend large sums of money to complete their collection, but consider the costs of measuring and analyzing their coins which are too high. The objective of the study [102] was to show, that worthwhile historical information can be obtained by analyzing old coins.

---
[2] From the Czech 'groš' and German Grosch.
[3] A Czech city listed and protected by UNESCO.

TABLE 10.4
Data support bounds and medians for coin weights

Type	Mintage Time	LB (g)	UB (g)	Median Med (g)
I.	1346–1348	3.233	6.369	3.478
II.	1348–1355	1.852	3.556	3.375
III.	1350–1358	1.458	3.590	3.402
IV.	1358–1378	0.201	3.723	3.297
V.	1370–1378	2.475	12.73	2.901

TABLE 10.5
Data support bounds and medians for purity of coins

Type	Mintage Time	LB	UB	Median Med
I–II	1346–1355	0.843	0.936	0.848
III	1350–1358	0.844	0.870	0.865
IV	1358–1378	0.837	0.865	0.863
V	1370–1378	0.643	0.767	0.763

Available Facts

The coins made available for analysis dated from the reign of the Czech King and Holy Roman Emperor Charles IV, (1346–1378). The designation of "types" and subsequent identification of the issue dates had been numismatically determined previously by comparing minor differences in the images, that had been struck. Tab. 10.3 reports the results of the analysis.

Robust estimates of lower and upper bounds of the data support (LB and UB) and median determined as quantile of probability 0.5 are reviewed in Tab. 10.4 for analysis of weights and in Tab. 10.5 for purity.

While modern purity measurement methods do not damage the coins, the traditional/historical metallurgical methodology for analysis was destructive, therefore the number of data that could be used in the cited study was relatively small. This explains the preference of collectors for untested and undamaged specimens over those of known purity, which have suffered under essay.

Distributions of Coins' Quality

The manufacturing technology for coins at that time was much the same, but cruder than what is common today. Purity was controlled by the amount of base metals included in the alloy, and a sheet was hammered out and rolled to a desired thickness by a mill. The coins were then stamped out using an engraved die. The master of the mint was responsible to the King for

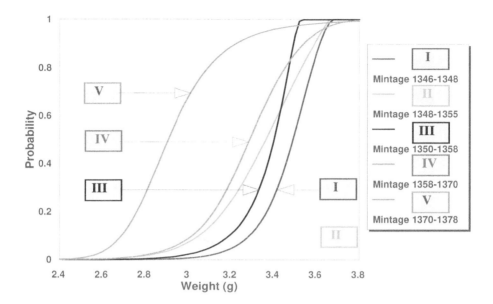

FIGURE 10.6
The effect of inflation demonstrated by movement of the distribution functions from right to left hand sides

controlling these two parameters; this was necessary for wide acceptance and trust in the coinage of the realm.

The relative spread for purity in Tab. 10.5 is much smaller than that of the weights. This can be easily explained by the fact that most of the coins were found in buried treasure, and corrosion had a greater impact on the coins' weight than on their chemical composition. A second likely reason is, that the ability to maintain standard weights over several centuries would have been more difficult than adhering to a purity norm.

The results show a gradual but substantial degradation in the value of the money. This conclusion can be supported and a more detailed insight into the process can be obtained by considering the distribution functions EGDF of weights in Fig. 10.6 and of the purity in Fig. 10.7.

The form of the distribution functions of the weight in Fig. 10.6 and the values of the upper bound of data support (UB) allow several conclusions to be drawn as to the development of the quality of production. The distribution of the oldest coins (type I) gives evidence of a very wide tolerance in the thickness of the silver sheet, which results in nearly random behavior of the function, and in high probabilities of exceeding the standard with a very high value of UB (6.369 g). The curve for the type II shows a substantial improvement.

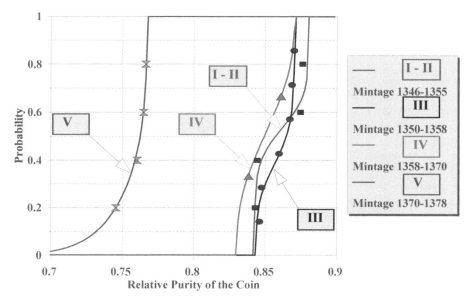

FIGURE 10.7
Deterioration of currency demonstrated by movement of the distribution functions from right to left side

But the shift of the distributions to the left side indicates deterioration of the money with time.

The standard (initial) weight was fixed and accurately maintained with $UB = 3.556$ g, to which the distribution function sharply rises.

Recall, that the steeper the probability curve rises, the more certain the quantiles are, and the less the "local" volatility. The curve for type III in Fig. 10.7 attracts attention from this point of view: it is the steepest one and—like the type II—it distinctly marks the upper break of the function close to the established standard (which can be estimated by the UB of 3.590 g). It is interesting to note, that according to historical records, there was a new and strict master of the mint appointed at that time. However, after 1358 the quality started to deteriorate with the UB of the type IV coins increasing to 3.373 g followed by a type V UB of 12.73 g: the "old good order of things" returned signified the decline of the society. This process can also be traced by values of the robust median, which was 3.402 g (type III) to fall to 3.297 g (type IV) and to 2.901 g (type V).

In the lower part of the distribution functions in Fig. 10.6 the spread of LB values is much larger than that of the UB. This is caused by deterioration through corrosion rather than by poor control over the weight of the coins.

Since the smallest values of the lower bounds belong to Types IV and V, this supports the comment on poor minting quality. This is further supported by Fig. 10.7, which shows, that in the case of type V the purity was poor.

The probability distribution functions of the coins' purity shown in Fig. 10.5 support the conclusions inferred from the weights. The bounds set out in Tab. 10.5 for types I, II, III and IV document good control for the coins' purity. The change in quality over time can also be observed: The initial upper bound UB was the highest for the types I–II (0.936). However, then it was decided to be more economical in the use of silver and to decrease the upper limit to 0.870 for type III, while keeping its lower limit unchanged ($LB = 0.844$), so as to maintain a solid foundation of precious metal. These results infer that a decision was made to lower the silver content, yet retaining quality to ensure, that the value of the coinage was maintained. The form of the functions here is even more unusual than in Fig. 10.6 instead of an S-form, the very tight bounds on purity force them to rise steeply and cut off sharply on both-sides of the bounds. Care in the quality of production seemed to be approximately maintained for type IV at the upper bound, but the lower bound LB was allowed to fall to 0.837. After 1370, all pretense for coin quality was abandoned completely: the upper bound UB of purity was drastically decreased to 0.767 and the respect for the reputation of the money obviously declined judging by the fall of the lower bound to $LB = 0.643$.

This brief look at the manufacture of coinage over a short period of history can provide some interesting insights in the development of society even though a significant period of time has passed, and the physical quantity of evidence is sparse.

10.7.2 Hypotheses Testing

Statistical testing of hypotheses is an important task which cannot be solved without distributions of probability because the decision on accepting or rejection of the tested hypothesis is based on probabilities. Recall the standard terms used in this procedure: **Null hypothesis:** The statement of zero or null difference that is directly tested (denoted as H_0). The finally conclusion is either rejection of H_0 or a failure to reject it. **Alternative hypothesis:** The statement that must be true if the null hypothesis is false. **Type I error:** The mistake to reject the null hypothesis when it is true. **Type II error:** The mistake of failing to reject the null hypothesis when it is false. α, **alpha:** Probability of a type I error. β, **beta:** Probability of a type II error.

The problem on practice is how to estimate the probability. To evaluate it by means of the frequency method can need a lot of experiments and data while finding a suitable distribution function can be difficult. The gnostic kernel estimated probability distribution can be applied, because it requires only some small data samples and does not need a priori assumption of the type of distribution. The first example of such an application is in Fig. 10.8.

There are two pairs of probability and density curves in this figure, one for the case of null hypothesis corresponding to the null medicine used against high blood pressure while the other corresponds to the alternative medicine. The null hypothesis is that there is no difference between the medicines. It can

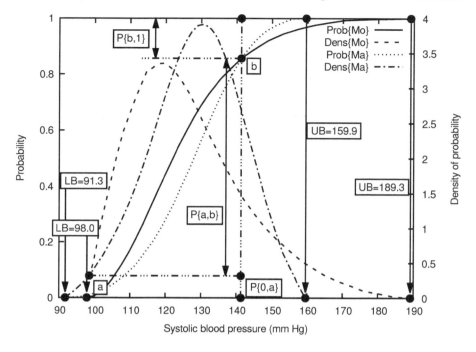

FIGURE 10.8
Testing the priority of a medicine

be seen that within a range of pressure from about 100 through 140 mmHg the effect of lowering the pressure of the null medicine is in the worst case about probability 0.2 stronger than that of the alternative. However, over the upper cross point of the curves at 142 mmHg the null medicine allows the pressure to rise till the point of probability 1 which correspond to the upper bound of probability support UB equal to 189.3 mmHg. The maximum blood pressure with alternative medicine is 159.9 mmHg. These results decided in favor of the alternative medicine although not quite in a standard statistical way. On the other hand, it helped to survive to one of the authors of the book because his blood pressure of 189 mmHg could be dangerous.

10.7.3 A Large Survey of Chemical Pollutants

Another example is connected with a large survey of impact of the chemical plant on the amount of permanent organic pollutants (POPs) in the fat of people living close to the plant. The survey was performed in 2003 by a team of the National Institute of Public Health, Prague, Faculty of Medicine, Charles University, Prague and Institute of Public Health Ostrava. Results were described in [103]. Data of the survey were entrusted to the authors of

Some Applications of Distribution Functions 97

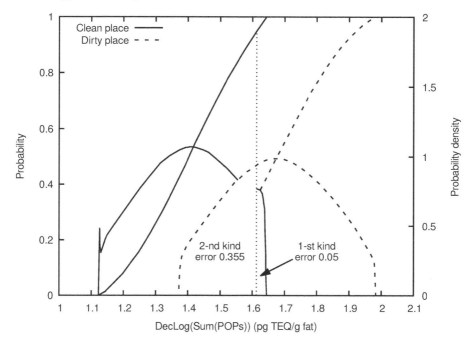

FIGURE 10.9
Testing the amount of contamination of two residences

the book thanks to prof. Černá from the Charles' University Prague to show the analysis not achievable by statistical methods.

It can be seen in the figure that all necessary for the statistical test of a hypothesis is easily obtained by crossing of the kernel estimates of probability distributions. More details of the contamination mechanism was possible by introducing an unusual parameter, *rate of accumulation of POPS* available as a ratio of the amount of POPS found in fat divided by the age of the tested individual. This parameter is based on the assumption that the contaminating process occurred during the whole life. Application of the local distribution function has not only supported the assumption but has also shown interesting details: the distribution obtained for men appeared to be non-homogeneous, revealing two sub-samples and two outliers 10.10:

The outlying values could be easily identified by means of finding details of the way of life of the two individuals: the lower outlying value was found at a lady which was used to consume vegetable from her garden watered by the water of the Elbe river. The extreme value belonged to a man consuming the fish he caught in the river. The homogenization of the two sub-samples of men enabled to study the interesting forms of distribution functions of both sub-samples 10.11.

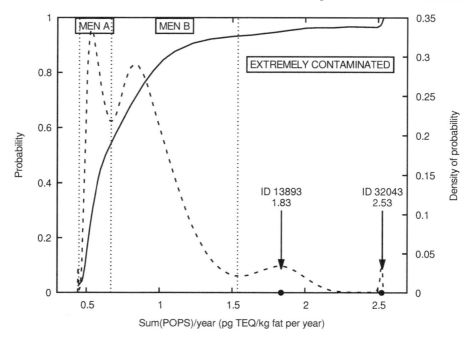

FIGURE 10.10
Local distribution function of the rate of contamination accumulation

The question of causes making the difference between two sub-samples of men can be answered by a look at the global distribution functions of both groups of men along with the distribution of women 10.12.

The higher rate of contamination of the sub-sample B of men could be caused by a longer time of being close to the plant because of working in the factory or in a place close to it. In contrast to it, the participants of the group A were working in Prague or in other places far from the plant: the time of being under the impact of contamination was significantly less. The distribution of women supports this explanation, because the probability of time of being "at home" was larger than that of men.

More significant information was obtained by means of multi-dimensional modeling. Two robust regression models were prepared analyzing the impacts on the total accumulation of POPs in the *dirty* and *clean* places. There were five variables influencing the accumulation:

1. distance from the considered chemical plant,
2. body-mass index of the participant,
3. rate of drinking alcohol,
4. rate of smoking,
5. the intercept the unexplained constant of the model.

Both models are shown in Figure 10.13.

The Homogeneity Problem

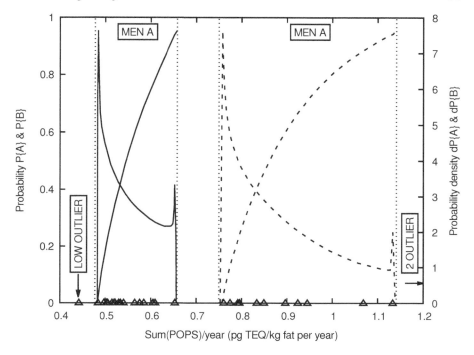

FIGURE 10.11
Local distribution function of two sub-samples of men

The dependent variable = the total accumulation of pollutants = appeared to be about two times large at the dirty place than in the clean place. The decisive impact decreasing the rate of accumulation in the clean place was the distance. It is not surprising that a comparable effect was that of body-mass index because the personal contamination was compensated by the weight. Drinking alcohol and smoking did not affect the accumulation of pollutants significantly. However a real shock was that of the intercept (the background value of accumulation having no explaining variable) which was practically the same independent of the distance from the plant. It is a bad message for all people living in the middle Europe: they are consuming a measurable amount of dangerous pollutants independent of their place of life.

10.8 The Homogeneity Problem

The problem of data homogeneity is one of cardinal ones in data analysis. The data sampling should reliably prepare data from the same family to ensure their mutual comparability and statistically treatable "mining" of their

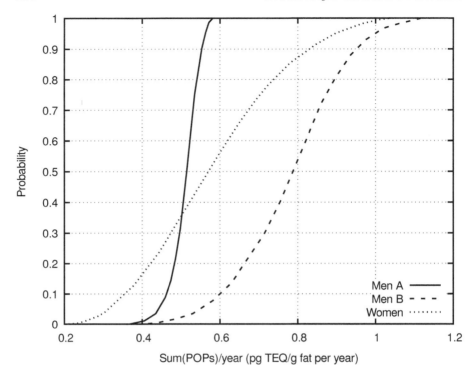

FIGURE 10.12
Local distribution function of two sub-samples of men

features. However, it is not an easy task. There are different causes for non-homogeneity, which are as follows:

1. There are outliers, data lying out of bounds of the "normal" data, caused for e.g. by disturbances which have nothing common with the process to be observed.
2. Instead of clean data of the same origin, a mixture of data from different sources are available.
3. The prescribed conditions of sampling were not respected.
4. The measuring system may be failing.
5. The required quiet working conditions during measuring process were disturbed.
6. The personal factors affected the measuring process.;
7. Some non-predicted factors entered into process, and
8. also, many other reasons.

The Homogeneity Problem

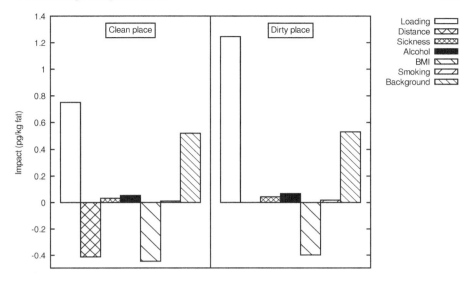

FIGURE 10.13
Impacts on the accumulation of POPs

A long-term effort in statistics has been devoted to elimination or at least minimization of such effects. However, experience of "data scandals" shows that the problems still exist. The hope is that creation of better data analysis methods would be helpful. This is why it was very pleasant to find that the global distribution function offers several intelligent ways to reveal the damage done to data from outside of the process: an outlier as well as a sub-sample causing the non-homogeneity reflect as a second maximum of the probability density. The algorithm looks for the largest maximum and its bounds and takes it for the main sub-cluster of the data. Then it finds bounds of the second (outlying) sub-cluster. Its data lying over the main cluster denotes by $+1$, the data below the main cluster by -1 and the data of the main sub-cluster receive the mark 0. Absolute values of these marks are summarized at the estimation end. If the result is zero, the sample is homogeneous, otherwise it is possible to eliminate the outliers by a program to get a homogeneous data sample. Functioning of this approach can be demonstrated on an example (Tab. 10.6). The matrix to be analyzed is called *swiss*. Column 1 of this matrix called Fertility characterizes the dependence of fertility measure for each of 47 French-speaking provinces of Switzerland at about 1888 on five socio-economic indicators (Agriculture, Examination, Education, Catholic, Infant mortality).

The column analyzed for homogeneity was column 2 (Agriculture). After application of the function $EGDF$, the control column appears in the form of Tab. 10.6.

There are two outliers reported: by indicator $Out = 1$ the upper outlier Herens and by indicator $Out = -1$ the lower outlier V. de Geneve. Looking

TABLE 10.6
The control column of the *EGDF* showing the outliers

Province	Out	Province	Out	Province	Out
Courtelary	0	Grandson	0	Herens	1
Delemont	0	Lausanne	0	Martigwy	0
Franches-Mnt	0	La Vallee	0	Monthey	0
Moutier	0	Lavaux	0	St Maurice	0
Neuveville	0	Morges	0	Sierre	0
Porrentruy	0	Moudon	0	Sion	0
Broye	0	Nyone	0	Boudry	0
Glane	0	Orbe	0	La Chauxdfnd	0
Gruyere	0	Oron	0	Le Locle	0
Sarine	0	Payerne	0	Neuchatel	0
Veveyse	0	Paysd'enhaut	0	Val de Ruz	0
Aigle	0	Rolle	0	ValdeTravers	0
Aubonne	0	Vevey	0	V. de Geneve	-1
Avenches	0	Yverdon	0	Rive Droite	0
Cossonay	0	Conthey	0	Rive Gauche	0
Echallens	0	Entremont	0		

into the data, we find that in province Herens 89.7% people were engaged in agriculture while in V. de Geneve only 1.2%. The statistical mean value of the Agriculture was Mean$\pm$STD $= 50.66\pm22.71$ covering the uncertainty interval from probability 0.186 through 0.830. The gnostic mean value was 55.41 ± 0.44 covering the probability interval from 0.513 through 0.527. The lower bound of the data support was estimated as practically zero ($7.49 \cdot 10^{-43}\%$) while the upper bound was 91.65%. The gnostic distribution function *EGDF* of the non-homogeneous data sample Agriculture is in Fig. 10.14.

The forms of distribution functions are disturbed by contribution of outliers. There are two possible ways how the non-homogeneous function can be homogenized:

A Homogenization by elimination.

B Homogenization by weights.

Elimination A is implemented by leaving out the outlier or outliers causing the non-homogeneity while preserving the other data. (The outliers are identified by indicators $+1$ or -1 in the control column in the resulting matrix of the *EGDF*.) This way is also eliminating by weights, because it gives weights 0 to outliers and 1 to other data. But the distribution function *EGDF* determines at each run the a posteriori data weights. These are applied instead of the a priori weights when using the method B. (The mean weight is kept equaling 1). There is no warranty that both methods will homogenize each matrix. In

The Homogeneity Problem

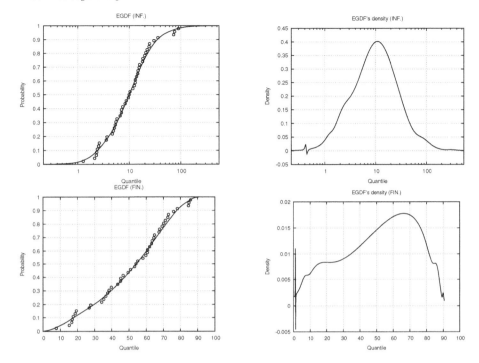

FIGURE 10.14
The non-homogeneous probability distribution function of variable Agriculture

the case of column Agriculture, the method A failed while eliminating all data, but the method B was able to prepare the homogenized matrix (Fig. 10.15).

There are interesting lessons from this case:

1. There are three factors determining that a data item becomes an outlier:

   data value: The extreme data will more probable become outliers than the other ones.

   data location: Data located close to the edges of the sample will be more probable to become outliers.

   data weight: The larger data weight, the larger the probability to become an outlier.

2. These factors play role on a very fine scale. The data enter the computing of the distribution function with the data with the same a priori weights 1. After finding the best fitting parameters of the probability distribution, the a posteriori weights are estimated as

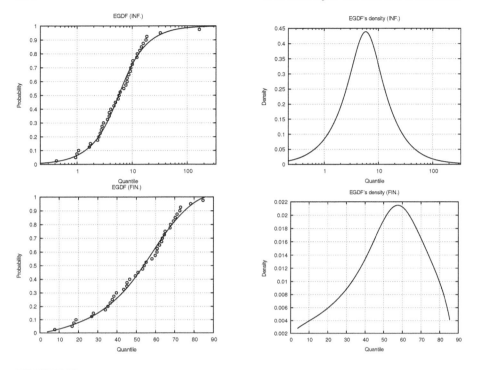

FIGURE 10.15
The homogenized probability distribution function of variable Agriculture

estimating data certainty with the mean value 1. Consider the extreme values of the a posteriori weights (Max. apow., Min. apow.):

- The non-homogeneous distribution: (1.088957511, 0.007418842)
- The homogeneous distribution: (1.088958077, 0.007418944)
- The extreme relative changes of weights caused by homogenization: (1.0000104, 0.9999863)

Such minute changes of a posteriori data weights were sufficient to make the non-homogeneous data sample *swiss[,2]* with its distribution 10.14 homogeneous with a nice smooth distribution 10.15.

10.9 Conclusions

There are four versions of kernel estimates of distributions differing by kinds (local and global) and by types (quantifying, estimating). They differ by their robustness with respect to outlying data and inner noise of data samples and by their flexibility. This determines them for different classes of applications. There are six categories of tasks solvable by the distribution functions reaching wide range of applications from the simple estimation the probability and density to deep insights into the structure of data samples. Being the application of the gnostic kernel (probability and density distribution of a single data item) they are not limited by the a priori assumptions on the standard statistical distribution and are usable for small samples of real strongly uncertain data. An example of a large survey on contamination of people has demonstrated that using the gnostic distribution functions opens the way to get more information from data than available by the statistical methods.

11
Applications of Local Distributions

The unlimited flexibility of the estimating local distribution functions *ELDF* in its dependence on the scale parameter S along with their unexpected behavior with respect to an additional data item makes them an ideal tool for a deep insight into the structure of a data sample. Following classes of applications of the *ELDF*s will be illustrated by examples in applications to real data from different fields:

1. Enrichment of the *EGDF*-analysis.
2. Revealing inner structure of a data sample.
3. Marginal analysis of a data sample.
4. Visualization of outliers.
5. Interval analysis.

Let us consider these classes in more detail.

11.1 Enrichment of the *EGDF*-Analysis

The estimating global distribution function (EGDF) results from an optimization process as an unique curve satisfying the given constrans to yield three of their parameters, LB, UB, S. It is robust (insensitive) not only with respect to outlying data but it also smoothes the small shifts of the inner sample's data. This makes its form very stable. Unlike this, by changing the scale parameter of the *ELDF* one can obtain details of the sample's structure which enrich the rough image produced by the global distribution. An example is in Fig. 11.1.

The object to be explored was the furnace of mark "Burner." The measured quantity was the Permanent Organic Pollutant (POPS) called 6CB (its molecular formula is $C_{19}H_{21}N$) in exhalation of the furnace. The sample treated by the *EGDF* appeared to be homogeneous, its density—as seen in the figure—had only one maximum. However, the *ELDF* has shown local clustering especially changing the distribution form at concentrations above 1.6 ng/m^3 N. The lecture for the explorer was: return to data to check, why this local change of the process originates and what measures should be made to prevent such rise in dangerous exhalations.

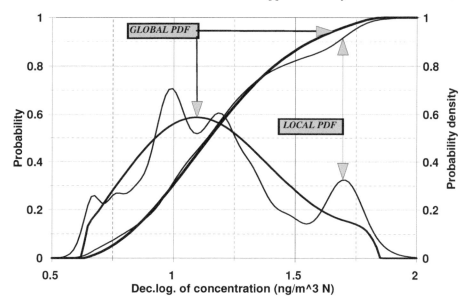

FIGURE 11.1
Comparison of global and local distribution

11.2 Revealing Inner Structure of a Data Sample

This example stems from a historical international measurement of features of the granite. In the beginning of 19th century a wave of interest to ward the investigation of granite started to increase in connection with hopes of shining the future of atomic energy. Granite was a perspective material for storages of burned atomic fuel especially by its capacity of absorbing thermal neutrons due to its cobalt content. This was why the international organization of geologists organized the comparative measurements of typical samples of granite. The treatment of data coming from a large number of world laboratories was entrusted to a Czechoslovak geological institution which had a test operation the in a gnostic analyzer. To their surprise the probability distribution function prepared by the *ELDF* appeared to have instead of one three distinct maxima.

The clustering was then explained by exploring the measuring methods and results of them fell into the individual clusters. Three classes of measuring methods were found giving biased results not corresponding to a true "mean" result. Publishing of this result in the world journal "The Analyst" arose the surprise worldwide [99].

This result (on different results with different methods) can be confirmed by the next example.

Marginal Analysis

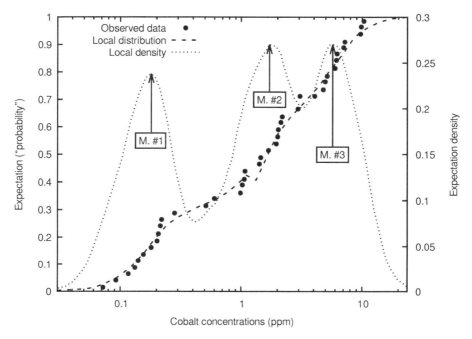

FIGURE 11.2
Non-homogeneity of measurements on granite

Three methods of measuring the concentrations of tantalum were compared: the mass spectroscopy, the activation analysis and roentgen-fluor spectroscopy. The results differed by their ranges and forms of probability density distributions. It is interesting, that medians of three measurements differed much less than the densities. It is an example of the difference between point estimates and distribution functions: the information obtained by the distribution functions is much richer.

11.3 Marginal Analysis

The homogeneity of an one-dimensional data sample is an important indicator of successful sampling allowing the data to be taken as a comparable one. Therefore the analysis should always start with homogeneity test by means of the *EGDF*. However, when the test proves the sample's non-homogeneity the question arises what is the cause and which data are contributing to the non-homogeneity. If they are not individual outliers, the marginal (one-dimensional) analysis may help to understand the "strange" distribution by

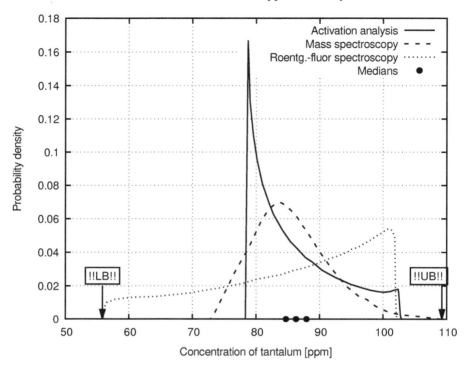

FIGURE 11.3
Comparison by three methods of measuring Tantal

means of the *ELDF*-kind with multiple local maxima of probability density. An example from the advanced financial statement analysis can demonstrate such application: denote *TEQ* as the total equity of a firm and *TA*, its total assets. The ratio *TA/TEQ* is called the *financial leverage* because it helps to increase the "financial power" of the firm because of increasing the total equity by an additional capital borrowed i.e. from banks. Using the financial leverage helps the financial management to invest more money into the productive activity and/or undertake some promising financial maneuvers. However, to decide the value of financial leverage is a serious economic problem because borrowing money is not free and the borrowed money must be repaid. This decision is therefore risky and it requires a greater look into "how our competitors solve the problem." Fortunately, the legal duty to make the financial statement periodically available enables public sharing the important data and to analyze them. The example is in Fig. 11.4.

The example shows the financial leverage of 13 firms of the big producers of Household Products Industry quoted on financial market of United States in 1993. The step function shows values of the weighted empirical distribution function. The light points correspond to halves of the steps and the smooth line the cumulative probability distribution *ELDF* with its density. There are

Marginal Analysis   111

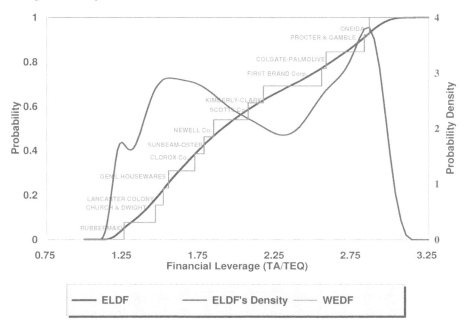

FIGURE 11.4
Marginal analysis of financial leverage

three maxima of the density function. The firms can be sorted into three clusters. The smallest financial leverage corresponds to the Rubbermaid Co. The second cluster includes firms from Church & Dwight Co. to Kimberly-Clark Co. The third cluster is formed by the four companies, First Brand Co., Colgate-Palmolive, Procter & Gamble and Oneida. The interpretation of clustering can be done by taking into account the size of firms and their relation to source of credits. The larger the firm and the broader is its dominating role in different countries, the larger can be the financial leverage without a danger of causing the financial instability. So, e. g. the two companies with the largest financial leverage (Colgate-Palmolive and Procter & Gamble) were running their branches in many countries of the world and keeping a dominating role in the market. This information is of course known to the financial manager and is used in making decision about opening an additional credit. The manager also knows the financial reserves of his enterprise and applies his financial strategy to balance the contradictory effects on economic gain and stability. It may be interesting to remember in this connection the time of Czechoslovak socialism when the maximum value of the financial ratio was ordered by law and when the financial manager could be fired for the value greater than the legal maximum and no economic factors were accepted as self-justification.

The marginal analysis includes following steps:

- The homogeneity test proving the sample's non-homogeneity.
- Experimental trials with the *ELDF* to find the value of the S distinctly showing the suitable number of separative maxima of the probability density function.
- Separation of data of individual clusters falling in between minima of the density function.
- Analysis of found clustering by using the data values and other related data to identify the factors causing the clustering.

11.4 Information Capability of Data

The principle "Let data speak for themselves" is a wise recommendation for statisticians who dwell on a-priori assumptions on data features and their a priori mathematical models. The mathematical gnostics takes care to keep this principle in all its operations. However, experience with thorough data analysis shows, that this principle is a little one-sided when letting the data speak for themselves (about processes they originated in). Factually, the information capability of data is far exceeding the requirement "to speak for themselves" being able to provide information "on other data" and "on other processes." This statement can be documented by another application of marginal analysis. The data of the example [86] shown as Fig. 11.5 stem from world statistics evaluating activities of individual countries in nuclear physics in 1993 measured by publication per 1 million of inhabitants.

This special publication activity started at 1 per million inhabitants in cases of Nigeria and Pakistan and finished about 500 per million in case of Luxemburg. The marginal analysis started with the scale parameter $S = 1.7$ appeared to show one maximum over the value of about 100 per million which separated the group of western countries ("like USA": USA, France, Great Britain, West Germany, Holland, ...) from the group of east and middle European countries ("like USSR": USSR, German Democratic Republic, Poland, Czechoslovakia, ...). This separation is clearly political, i.e. socialist versus capitalist countries. Moreover, when a lower scale parameter $S = 1$ was applied, two clusters appeared separated by a minimum exactly corresponding to the criterion of the United Nation Organization distinguishing the developing countries from the developed ones. This is an economical and demographical information. Further diminished scale parameter $S = 0.22$ made the political information more detailed: two countries from the group "like USSR" were separated as

Interval Analysis 113

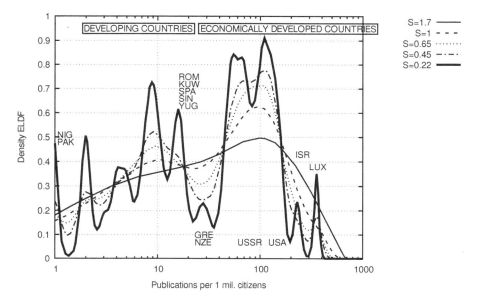

FIGURE 11.5
Marginal analysis of publication activity in nuclear physics

non-socialist, Greece and New Zealand and Romania with Yugoslavia fall under the group "like USSR" in spite of their socialist economy but making politics not exactly like USSR. Two clear outliers separated in the group "like USA," Israel and Luxembourg. In the case of Israel the cause is clear, as atomic weapon is necessary for its survival in Arabian encirclement. Interpretation of leadership of Luxembourg would require extension of the information by historical facts from the scientific development of the country. Anyway, the example shows that data about scientific publications are capable to reveal information about geographic, demographic, political and even military character and allow to get an insight even the details of power politics.

11.5 Interval Analysis

The robustness of *EGDF* is connected with the low flexibility of this distribution function. Unlike *EGDF*, the local curvature of function *ELDF* is decreasing unlimitedly with the scale parameter's value. An impression, that such a high flexibility results in a non-robust behavior, would be erroneous. This can be shown by an experiment:

1. Take a homogeneous data sample $\mathcal{Z}$ of N data and run the *EGDF* to get the best estimates of the bounds *LB*, *UB* and of the scale parameter S.

2. Run the *ELDF* by using the same parameters *LB*, *UB* and *S* and find the location ($\tilde{Z}_{0,N}$) of the (unique) maximal density (the mode) as an estimate of the true value Z_0.
3. Extend the data sample $\mathcal{Z}$ by an $N+1$-th item z and find a new mode $\tilde{Z}_{0,N+1}$.
4. Let the z successively rise from a very small to very large values (while leaving the data of $\mathcal{Z}$ and S fixed) registering the values of the function $\tilde{Z}_{0,N+1}(z)$.

Behavior of this function is unexpected revealing five remarkable points Z_L, $Z_{0,L}$, Z_0, $Z_{0,U}$ and Z_U:

- In the point $z = \tilde{Z}_{0,N}$ the equivalence $\tilde{Z}_{0,N} = \tilde{Z}_{0,N+1}(z)$ is reached.

- For extreme values $z \to 0$ and $z \to \infty$ the limits $\lim_{z \to 0} \tilde{Z}_{0,N+1}(z)$ and $\lim_{z \to \infty} \tilde{Z}_{0,N+1}(z)$ are the same equaling $\tilde{Z}_{0,N+1}(\tilde{Z}_{0,N})$.

- Over the interval (LB, Z_L) the function $\tilde{Z}_0(z)$ uniformly **falls** to reach its local minimum $Z_{0,L} = \tilde{Z}_{0,N+1}(Z_L)$.

- Over the interval (Z_L, Z_U) the function $\tilde{Z}_0(z)$ uniformly **rises** to reach its local maximum $Z_{0,U} = \tilde{Z}_{0,N+1}(Z_U)$.

- Over the interval (Z_U, UB) the function $\tilde{Z}_0(z)$ uniformly **falls** to reach its limit $\tilde{Z}_{0,N}$.

Only the behavior over the interval (Z_L, Z_U) can be called *typical*, because increasing the extending data item z leads to increasing the location parameter $\tilde{Z}_{0,N+1}(z)$[1]. Behavior over the interval LB, Z_L as well as over the interval Z_U, UB are **atypical**. There is a good reason to call these intervals *lower atypical* and *upper atypical*, correspondingly.

The interval $(Z_{0,L}, Z_{0,U})$ also deserves attention: the location parameter $\tilde{Z}_{0,N+1}(z)$ cannot leave this interval, when z passes through the infinite interval $(0, \infty)$. It can be called *the tolerance interval*, because it defines the location parameter's tolerance under all possible changes of z. The length of the tolerance interval is obviously smaller than that of the interval (Z_L, Z_U), but its finiteness is sufficient to support the statement, that the *ELDF*'s location parameter is robust.

Two *membership bounds*, the lower and upper can also be obtained as quantiles corresponding to points of inflexion *LSB* and *USB* of the probability density function. They delimit the domain of data that are members of a homogeneous sample.

The mentioned points are defined mathematically [62] and they can be computed in a process called *interval analysis*. This analysis enables data values D of a homogeneous sample to be classified as belonging to one of nine

---

[1] The notion of typicality was inspired by behavior of a data mean under a similar situation.

classes of intervals of the extending variable Zx and of the variable $Z(Zx)$ (the variable location parameter). Intervals of Zx:

Lower improbable: $0 < D \leq LSB$

Lower outlier interval: $LSB < D \leq LB$

Lower atypical: $LB \leq D < Z_L$

Lower typical: $Z_L \leq D < \tilde{Z}_{0,N}$

Location parameter: $D = \tilde{Z}_{0,N}$

Upper typical: $\tilde{Z}_{0,N} < D \leq Z_U$

Upper atypical: $Z_U < D \leq USB$

Upper outlier interval: $USB < D \leq UB$

Upper improbable: $UB < D < \infty$

Intervals of $Z(Zx)$:

Lower tolerated: $Z_{0,L} \leq Z(Zx) < \tilde{Z}_{0,N}$

Location parameter: $Z(Zx) = \tilde{Z}_{0,N}$

Upper tolerated: $\tilde{Z}_{0,N} < Z(Zx) \leq Z_{0,U}$

The constant $\tilde{Z}_{0,N}$ is the location parameter of the fixed sample. An example obtained by application of the interval analysis to real data is shown in Fig. 11.6.

The interval analysis provides thus a very fine and reliable insight into the structure of a data sample. Its power results from the natural features of a data sample and not from some subjective assumptions on their features. This type of analysis can be used to objectively classify the data by their membership in sub-intervals.

11.6 Diversity of Samples

Samples of uncertain data can differ even in the case of measuring the same object or process. It is sometimes required to evaluate the degree of their diversity. Such an evaluation can be achieved by the interval analysis of data samples and by comparison of results. Take two data samples A and B and consider results of their interval analysis obtained by application of function *Gintervals*. Symbols ∩ will be used for disjunction of intervals and symbol 0 for the empty interval. The similarity of the samples can be classified in the following way:

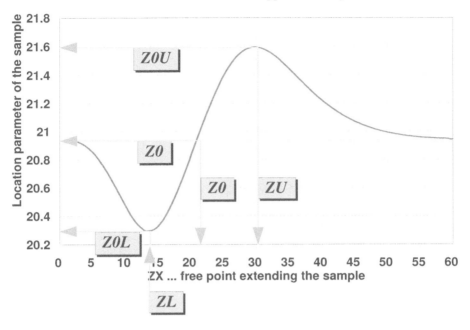

FIGURE 11.6
Interval analysis of pollutants PCB28-31

I. Considerable similarity: $\tilde{Z}_{0,A} = \tilde{Z}_{0,B}$.

II. Tolerated dissimilarity: $[Z_{0,L,A}, Z_{0,U,A}] \cap [Z_{0,L,B}, Z_{0,U,B}] \neq 0$.

III. Similarity of typical values: $[Z_{L,A}, Z_{U,A}] \cap [Z_{L,B}, Z_{U,B}] \neq 0$.

IV. Similarity of probable values: $(LB_A, UB_A) \cap (LB_B, UB_B) \neq 0$.

V. Improbable similarity: $(LB_A, UB_A) \cap (LB_B, UB_B) \equiv 0$.

The similarity of the first class is achieved by equality of location parameters although the data samples are not identical, but they provide to the analyst the same estimate of the true value (although with different weights). In the case of at least partial covering of toleration intervals (class II) it is possible to find an extending data item (the same for both samples), so that the location parameters coincide. The diversity of the samples may be tolerated. Similarity to the class III is only of a partial significance, while the class IV says only that the similarity cannot be completely negated.

This analysis may be used in comparison of quality of different laboratories. Such comparisons are very popular and their surveys are used for different purposes.

The diversity degree can be quantified by determining the probability attached to individual intervals.

11.7 Conclusions

The flexibility of the Local Estimation Distribution function in dependence on the scale parameter along with its internal robustness allows a large variety of tasks to be performed:

- Yielding additional information to the analysis performed by the global analysis,
- investigating the data sample's structure,
- implementing the marginal analysis of a data sample and one-dimensional cluster analysis,
- revealing indirect information on processes by making use of the information capability of data due to the universal interdependence of real processes,
- performing the robust interval analysis to get the objective bounds of several intervals,
- establishing the similarity of different data samples by means of the interval analysis.

12
On the Notion of Normality

12.1 Normality of Data

The notion of normality is one of the most important characteristics of life. It is inherently connected with decision making because deviation from the normal state or process requires correcting or control actions. There are two approaches to the determination of such a deviation:

1. Qualitative—by comparing the actual symptoms of the state or process with those generally accepted as normal.
2. Quantitative—by measuring some decisive parameters and checking if their values lie within the normal (reference, tolerance, alarm) bounds.

Only the Quantitative way will be dealt with here.

The statistical notion of normality is strictly bounded with the Gaussian distribution because of many advantages of this special distribution. However, there exists much more "named" distributions in statistics and in other scientific fields and a lot of non-named distributions: each measurement gives rise to a distribution. This book does not presume some special data and therefore will apply the notion of normality in agreement with the dictionary: normal is natural, usual, ordinary, conforming with an accepted standard, average, what you expect.

An inevitable problem in making the determination of the bounds of "normal" activity difficult is uncertainty. The measurement of real quantities is always disturbed by unknown factors. Data obtained by measuring are therefore contaminated by uncertain components. The bounds of normality can therefore only be obtained through processing the data based on a theory of uncertain events. A seemingly "trivial" solution, the application of mathematical statistics, is far from ideal. Statistical models of real processes and their corresponding measuring errors are not always available. The ever changing and complex nature of reality make such models unreliable. Moreover, classical statistical methods are not sufficiently robust. Many methods of robust statistics exist, but they are not universally applicable. The selection of the "right" method from the variety of statistical methods suitable to a particular problem can thus require a research task on its own. The statistical requirement for the availability of a "sufficiently large" data set to solve the problem does not always correspond to real conditions of measuring. The empirical way frequently used in certain fields—especially medicine—suffers from a lack of universal applicability and reliability. There also are methods based rather on intuition or subjective impressions than on data. It is worth of having a look at some of these methods.

12.1.1 Statistical Approach

A given set of random data with a given probability distribution of the set and a given significance level is considered. The role of the lower bound of the reference range is assigned to the quantile, the probability of which corresponds to the significance level. The upper bound is the quantile of the probability one minus the significance. A value randomly taken from the set belongs to the reference range if and only if it lies between its bounds. This approach seriously suffers from at least three degrees of subjectivity:

1. The probability distribution of a real data sample is rarely a priori available. Its statistical model must mostly be assumed. The assumption depends on skill of the person doing the analysis and on specifics of data. A frequently used assumption to use Gaussian or log-Gaussian distribution is not always realistic especially due to the mismatch of the finiteness of real quantities and the infinite Gaussian domain and of having a density form differing from the bell form. In addition, the zero value acceptable from the Gaussian point of view cannot model a realistic, usable observed value. The assumption of "Gaussian normality" is also hidden in *ISO* norms requiring the data mean and a $\pm K$-multiple of the standard deviation to play the role of the reference bounds. The "normalization" of non-Gaussian distributions by means of data transformation followed by tests of the "normality" of the resulting distribution cannot be a proper remedy because the transformation introduces additional, artificial data interdependencies, which can reduce the reliability of the decisions based on these data.

2. Statistical manuals require an a priori setting of the significance, normally performed before gathering the data. However, experience shows that if a statistical test does not confirm the expectations or more importantly the requirements of an experimenter or analyst, the change of the "required" significance is applied to make the experiment successful. This then leads to the argument that these choices are arbitrary and will further degrade the quality of the decision making by this method.

3. Relying on two statistical moments, mean and standard deviation is also based on the Gaussian assumption and on its preservation. These statistics are efficient in the Gaussian case (sufficient and

necessary for estimation of the distribution) but not in cases of modified distributions. Changes in the distribution form result in changes of probabilities and risks.

Moreover, the non-robust nature of the statistical moments can substantially increase the risk of incorrect decision making. Unfortunately, this approach is not only very popular on practice but also prescribed by some norms. It can thus happen that the *ISO*-conforming product or process is declared as normal although it is far from "ordinary or what you expect" as a dictionary requires from the word. There are data analysts who never met undoubtedly normal (Gaussian) data. Another problems are caused by false data aggregation generally applied and prescribed by many norms.

12.1.2 Empirical Way in Clinical Practice

Quantitatively supported diagnostics is based on *reference values*, established by clinical practice in the empirical way. Reference values are defined as bounds of a range of a majority of values obtained by measuring the reference population. The "reference population" is composed from individuals not suffering by the disease, which is to be determined. "Majority" is usually 95%. This definition is vague because of difficulties of selecting the reference ("healthy," "normal") population. The health state of individuals is dependent on many factors, and therefore it is a multidimensional object. The "normality" of the reference population is thus questionable and must be subjected to a reliable test. Application of the reference values to other individuals requires the ability to compare tested individual with the reference population, which is not fully comparable. Moreover, the "cutting of" the 5% the reference population is illogical and impractical:

1. These individuals are originally accepted as "healthy" to be later rejected because of their outlying ("non-healthy") parameters. Application of the "95%-reference values" to testing the "healthy" individuals may cause a false alarm in 5% of cases.

2. Bounds of the normal range must be broader than the interval between minimum and maximum observed values, because even the extreme "peripheral" but "normal" observed values are uncertain. Therefore, the probability of another "right" observation lying out of this interval cannot be zero. Determination of reference/normal range of diagnostic parameters should be based on arguments, which would be more rational than the intuitive ones recently being applied.

3. The homogeneity of the reference sample is not tested, especially because there is a lack of a proper statistical method.

4. The method requires a lot of "healthy" patients for the control group which may be not easily available.

The bounds of the normality/reference range serve to diagnostic decisions and to intervention into the patient's health state. The subjective of accepting the size of the "majority" of 95% also determines the risk of a bad decision making of the physician.

Using gnostic distribution function for evaluation of reference values could enhance the reliability and objectivity of this important medical method and improve its economic efficiency because of requiring less participants.

12.1.3 Similarity-Based Reference Values in Economy

Decision making in the financial management of business and investment activities is dependent on "measuring" the performances of economic objects, which is difficult due to its multi-dimensionality, non-stationary nature, as well as, the uncertainty resulting from the impact of individual participants following their personal or group interests. The traditional "measuring instrument" of economic activity is the financial statement analysis and rating [24]. These are far from the perfect method in providing reliable insight into economic processes as shown in detail in [81]. The traditional approach consists of deriving a set of ratios of basic financial parameters of business enterprises and comparing them with some broadly accepted "recommended" values obtained by averaging these ratios over the industry. The state of multidimensional objects is thus characterized by a multiplicity of one-dimensional parameters. Application of these ratios instead of the absolute values of the quantities used in the ratio is motivated by the necessity of removing the size factor, i.e. to make small firms comparable with the large ones. This approach suffers from several flaws:

- The classification of firms from the point of view of their membership in an industry does not warrant economic comparability of all the businesses in the group. Experience shows that some members of an industry cannot be compared with others because of differences in technology, clientele, raw materials used, geographic location, market orientation, as well as other factors. On the contrary, economically comparable members of different industries exist. The comparability of firms still remains a problem.

- The deviation of a ratio from the "reference value" can be compensated by deviation of another ratio of a multidimensional object without disturbing the object's "normality" and therefore multidimensional analysis is a must. The ordering of multidimensional objects cannot be based on consideration of some individual dimensions and represents a complex problem. The same relates to the similarity of such objects.

- It is impossible for humans to perceive all the multidimensional events within the decision process. The control of multidimensional objects requires mathematical models to be applied.

Both traditional and advanced financial statement analysis are discussed in [81].

Normality of Data 123

12.1.4 Fuzzy-Set Approach

The membership problem in classical (Cantor) set theory is solved by introducing a primitive assumption[1]: everybody knows when x is a member of a classical set X. However, fuzzy-set theory introduces the membership function evaluating the grade of membership of an element x in a fuzzy set X. This concept is more realistic than the "purely mathematical" one because it takes into account uncertainty. On the other hand, identification of such a model using real data represents a problem. The simplest fuzzy models apply triangular or trapezoidal form of the membership function defined by 3 and/or 4 points of the data support. Consider the trapezoidal form of a membership function standardly assumed:

This function consists of a "left-hand" triangle (LB, LC, A), of a square (LC, AC, BC, UC) called *core* and of a "right-hand" triangle (LC, BC, UB). The points LB, LC, UC and UB lie on the horizontal x-axis along with the fuzzy-set's data. The interval between points LB and UB is the support of data. Data values from the interval between LC and UC are definite members of the fuzzy set, their grade of membership (μ) is 1. The grades of membership (μ) in the outer intervals increase/decrease from 1 to 0. Fuzzy set is called normal if its height (the maximum of the membership function represented by the horizontal (AC, BC) is 1. Its subset of values which have a grade of 1 is the core. This concept is realistic because:

1. The finite nature of real quantities (limited by the lower and upper bounds of interval (LB, UB)) is respected.

2. Inevitable uncertainty of knowledge on quantitative events is taken into account, characterized by intervals (LB, LC) and (UC, UB).

3. The existence of a real core of the set, i.e. the definite members of the set) is described by the interval (LC, UC).

The identification of such a model using real data can be difficult even in this simplest form, let alone when additional questions are posed, such as:

A How do you realistically and robustly estimate the tetrad of decisive points on the x-axis?

B Does the linear form of the functions over (LB, LC) and (UC, UB) characterize the distribution of the data properly? What form should this take to correspond to the data uncertainty?

C Is the role (weight) of all the core data in the determination of the bounds LC and UC the same as a result from their full membership grade?

D What is the form of the density of core data corresponding to local densities of the set?

---
[1] Primitive is an assumption accepted without a proof.

There exist a lot of "manuals," as well as automatic methods of estimation of the fuzzy-set model. The manual methods based mainly on expert opinions are generally considered to rely on very subjective interpretations of words dependent on the vagaries of human experts and on knowledge acquisition problems. Many reported methods of automatic estimation, e.g. artificial neural networks, genetic algorithms, deformable prototypes, gradient search methods and inductive reasoning, are not supported by a consistent mathematic theory of uncertainty. Rather, they can be thought of as individual inventions solving particular tasks without the prospect for universal application.

The described form of the membership function can be thought of as an approximative model of probability density. If so, then it only stays to ask, why not use a gnostic density function determined by data.

12.1.5 Automatic Warning and Emergency Systems

The current level of world industrialization is managed by the mass application of automatic devices to continuously monitoring the maintenance of the normal state of life and technology. The bounds of normal ranges of monitored parameters are typically set by the previously discussed statistical methods, the weaknesses of which have been exposed. The rapid development of microcomputers and sensor technology makes the idea of the integration of sensors with signal treatment and communication elements into distributed network of compact monitoring devices feasible. The continued increase of requirements for the continuous control of environmental parameters drives the application of such autonomous devices which are placed on all chimneys and water outlets of manufacturers which generate dangerous pollutants, as well as, useful goods. A network of such devices would include the computers of all the institutions and firms engaged in the continuous care of the environment. High reliability, maximum information yield and adaptivity to changing situations would be a must for such systems.

12.2 Requirements to Ideal Estimation of Bounds of Normality

The discussion of the approaches analyzed above can be summarized in the form of the list of requirements for the applied method used to solve quantitative problems of normality:

1. The model used for decision making must draw out all necessary information from the data objectively, without any special a priori assumptions on the data model.

2. The maximum amount of information to be drawn out from the data must be proven by a consistent theory of the treatment.
3. The bounds of the domain of probable data must be robustly estimated from this data.
4. A mathematical definition of the range of normal data as a sub-interval of the probable data interval must be available and its bounds estimable from the data.
5. The method of estimating the bounds of probable and normal members of one-dimensional data samples must be also applicable to multidimensional samples.
6. The method must allow the option between two mutually opposite kinds of robustness of data treatment:

   - Robustness with respect to outlying data (outer robustness) and
   - Robustness with respect to inner disturbances/noises of the data sample (inner robustness).

7. The method must be applicable to data samples formed as mix of two or more sub-samples containing different ranges of normal data.
8. A method satisfying these requirements must be available in the form of programs run on generally available computing environments.

12.3 Elements of Gnostic Solution of the Normality Problem in a One-Dimensional Analysis

There are four versions of the probability distribution functions mentioned above. There is a notable feature of the function $EGDF$ which enables the robust testing of data homogeneity: the $EGDF$'s density function of a homogeneous data sample is uni-modal. The extension of such a sample by an additional data item reaching some extreme values leads to the occurrence of another maximum of probability density transformed onto the semi-infinite data support. At the "critical" value of the item, the point of inflection is reached. There are two such points denoted LSB and USB called *bounds of membership*. They can be also estimated after the $EGDF$ is made available. Note that these bounds are unique and objective, determined only by data, without any assumptions on their statistical features or on the significance of the test. The probabilities of reaching these values are easily calculated because the probability distribution is on hand. The question of data homogeneity is answered easily by the values of the result of analysis. This enables following definitions and procedures to be accepted:

Procedure A (estimation): Given a vector S of uncertain data ("one-dimensional data sample"). Run function *EGDF*.

Definition 1: A single-dimensional sample/vector of real uncertain data is homogeneous if its *EGDF* density over the infinite support has only one maximum. The test of homogeneity of a one-dimensional sample consists in checking that there is only one probability density maximum or—in the case of a density of the U-form—just two density maxima.

Definition 2: The bounds of normality of a sample S satisfying the definition 1 are identified by the points *LSB* and *USB* determined by estimation.

Procedure B (homogenization H): Given a non-homogeneous data sample S and its global gnostic distribution *EGDF*.

Alternative HW: The alternative HW achieves homogeneity of the sample S by giving to its components a posteriori weights instead of their a priori weights applied to the call of *EGDF*. This is iteratively repeated until the stabilization of the weights or until reaching the limit of repetitions. (Not all non-homogeneous data samples can be homogenized in this way). The advantage of the alternative HW is the preservation of all the components of S as no data are lost.

Alternative HE: This alternative (usable always) eliminates the elements of the vector S, which are signed by the results of the *EGDF* as causing the non-homogeneity.

The estimating global distribution function *EGDF* is based on the idea, that if a data sample is homogeneous, then it is possible to represent it by a single gnostic event (by a single equivalent data item), the estimating weight and irrelevance of which can be calculated by normalization of their arithmetical means by the estimating modulus N_I (10.5).

Checking of normality by means of gnostic algorithms can be implemented by a lot of ways:

1. By crossing a given quantile: There are homogeneous kernel probability distribution function available, therefore it is easy to implement a test for the required robustness and required quantile.
2. By means of the bounds LB and UB of data support.
3. By means of robust testing of hypotheses.
4. By means of classification of the danger by interval analysis.
5. By means of classification of the danger by data calibration.

The ordinary statistical way of controlling the normal state of things is to monitor the control bound set to mean value of the controlled quantity plus an addition of k-times the standard deviation. Application of this control to periodic measurements of a real process (control of exhalations of PCB28-31, chapter Homo- and heteroscedastic data) shows that when trying the k equal to 3 and 2 (values frequently applied) the control is missing disturbances which really occurred and which exceed the gnostic bounds set to the gnostic value *USB*.

12.4 Critics on the Identity Gaussian ≡ Normal

There is a fundamental objection against considering the Gaussian distribution "normal": the infiniteness of the Gaussian distribution is in a conflict with finiteness of real quantities subjected to measuring. Such an identification leads to problems in modeling the events with extreme probability (too small and too large probable events). Another problem is the symmetry of the distribution: why should be the positive deviations from the mean similarly distributed like the negative ones? Yes, the Gaussian applied as an approximation of an actual data distribution is theoretically advantageous, but is it sufficient for ignoring the miss-match with the real data? The mathematical gnostics respects the data and does not work with assumptions on their distribution, it simply uses the distribution estimated by means of the data as they are. The relation to the notion of normality is similar: normal events are those satisfying some conditions estimated by the data. The normality defined in this way includes not only the data values ("to be expected"), but also their uncertainty which is also "to be expected." An example may be the normality bounds of contamination of Moravian rivers Fig. 12.1.

Note that all bounds are finite and different by pollutants.

12.4.1 Re-definition of Normality

The possibility to quantify the amount of uncertainty enables to introduce a new notion of data normality. Take a data sample D and its k-th data item. Denote

- U_0 <- Contribs(D)[[1]][[11]][k]

- U_k <- Contribs(D)[[1]][[11]][-k]

where U_k is the sample U_0 with the k-th item left out. **Definition**: The k-th data item is ***normal*** iff $U_0 - U\_k < 0$, otherwise it is an ***outlier***. The value of k-th data item is thus triplified, split into three parts: ($|theloweroutlier|normalvalue|theupperoutlier|$). The bounds of the domain of the normal data items LB and UB are thus obtained as well as the smallest and largest normal values. However, more is obtained: the amount of uncertainty of each sample's data item. It is done by the function Gmean called

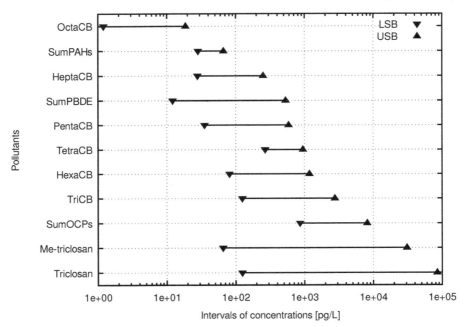

FIGURE 12.1
The gnostic bounds *LSB* and *USB* of Moravian rivers

repeatably by the function Contribs.[2] By running the function Gmean, one obtains the evaluation of the sample's uncertainty as the ratio $(f_J - f_I)/(f_J + f_I)$.

Example of triplication of the sample of contamination of Moravian rivers by the pollutant TriCB:

```
$HOM
        k  D[k]  Contrib.to Unc.
 [1,]  28   212   -0.000404
 [2,]  26   242   -0.001749
 [3,]  11   254   -0.002224
 [4,]  12   367   -0.005458
 [5,]  37   424   -0.006462
 [6,]   1   432   -0.006576
 [7,]  39   489   -0.007224
 [8,]  30   493   -0.007259
 [9,]  33   520   -0.007464
[10,]  32   641   -0.007792
[11,]  17   708   -0.007643
[12,]  19   714   -0.007621
[13,]   5   727   -0.007567
```

---

[2] Both functions Contribs and Gmean can be obtained by asking the publishing house.

```
[14,] 36  731   -0.007550
[15,] 35  738   -0.007518
[16,] 23  760   -0.007406
[17,] 38  764   -0.007384
[18,] 34  893   -0.006455
[19,] 15  909   -0.006315
[20,]  4  913   -0.006280
[21,] 31  945   -0.005987
[22,]  3  950   -0.005940
[23,]  2 1101   -0.004400
[24,] 21 1220   -0.003088
[25,] 25 1232   -0.002953
[26,] 18 1252   -0.002728
[27,] 22 1340   -0.001733
[28,] 14 1386   -0.001211

$LOUT
    k D[k] Contrib.to Unc.
1   7   32    0.039426479
2   8   61    0.018140262
3   9   35    0.035476160
4  10   33    0.038022135
5  13  134    0.004844652
6  27  126    0.005639262
7  29  175    0.001667905

$UOUT
    k D[k]  Contrib.to Unc.
1   6 2682    0.0123914828
2  16 2357    0.0091936122
3  20 1573    0.0008982297
4  24 1629    0.0015222654
```

The matrix $HOM is formed by the normal data, $LOUT by the lower and $UOUT by the upper outliers. The k is a data serial number in the sample, the second column shows the order of the item and the third column are names of the data items. The forth column are contributions of individual items to the sample-s uncertainty.

12.4.2 On a Still Daydreamed Research Project BONUS

There was a nearly ten years lasting period of rich and long history of development of the mathematical gnostics of a collaboration with the Institute of Public Health in Ostrava oriented on the application of new data analytic methods to the problems of environmental topics. It was especially welcome accepting this chance which was made possible by understanding of the specialists of the Institute because of the following:

1. This Institute works as the reference laboratory for research and control of Persistent Organic Pollutants (POPs) and of contamination of the environment and food by POPs.
2. Measuring of this contamination is very difficult, because only some trace amounts of POPs are sufficient to jeopardize the people's health.
3. These measurement are expensive and data are therefore available only in small amounts.
4. The problems of environmental care are international and the Institute was collaborating with the French, Italian, Polish and German partners in the framework of research grants of the European Union.

Gnostic methodology has been successfully applied in a large number of cases proving its adequacy to solution of the specific environmental problems. It was therefore logical to think of a follow-up project to further develop the analytic technology. There was even a suitable acronym prepared for the project: BONUS (Bounds Of Normality of Uncertain Situations). The basic idea of necessity of such a project resulted from an analysis of data of a survey accomplished by the Institute of Endocrinology, Prague. The survey followed the recent tendency in health care: the medicine should become to be **personalized**[3]. Everything starts with diagnostics. The ideal diagnostics would take into account as much factors as possible to use personal diagnostic bounds established for individual patient instead of the "collective" bounds. To demonstrate the effect of personal bounds of normality, the multidimensional endocrinologic data measured in the Endocrinologic Institute were applied to find gnostic "bounds of normality" of estradiol of a group of 114 postmenopausal women. There were 33 parameters considered in the study including 8 genes with 3 alleles. There were thus 8 genetic factors selected from all 24 defining the state of each patient. Quantification of composition of these factors enabled the gnostic bounds of normality of the estradiol in considered women to be estimated. It was found that the data were drastically non-homogeneous. A detailed analysis of factors causing the non-homogeneity has shown, that the normality bounds were strongly dependent on genetic factors. Data were therefore split into 24 sub-samples of women with the identic genotype. Intervals between the lower and upper gnostic bounds of normality of each sub-sample are shown in Fig. 12.2 by horizontal lines. The recently used official reference values of estradiol are presented by two vertical lines.

The results show that the individual diagnostic bounds strongly depend on genetic factors: what amount of estradiol is "good" for a woman can be

---
[3]The data were made available to one of the authors by Prof. I. Žofková to whom we obliged to be thankful.

Critics on the Identity Gaussian ≡ Normal 131

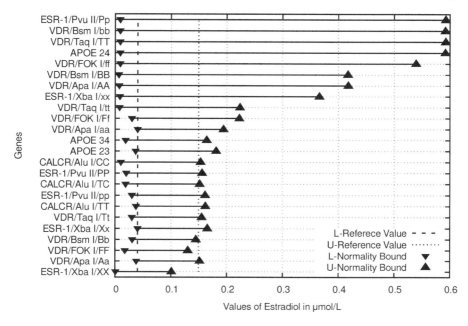

FIGURE 12.2
Actual bounds of estradiol in women

"wrong" for another woman differing in structure of genetic factors. The differences between individual reference bounds are extraordinarily large, the higher bound was about four times exceeding the official maximum reference value in four sub-groups. The official maximum was exceeded by a halve of 24 groups. Only imagine that your mother or wife would be operated or otherwise drastically "medicated" because of a fault reference value! Such a feeling was motivating writing and submitting the project to the grant leadership. The project included the following topics:

1. Introduction
2. Basic approaches to estimating the bounds of normal/reference range

   [2.1] Statistical approach

   [2.2] Empirical way in clinical practice

   [2.3] Similarity-based reference values in economy

   [2.4] Fuzzy-set approach

   [2.5] Automatic warning and emergency systems

3. Requirements to ideal estimation of bounds of normality
4. Feasibility of the project

[4.1] Available approach to the solution: mathematical gnostics

[4.2] Potential elements of gnostic solution of the problem
- One-dimensional data sample: Function GNDF
- Multidimensional data sample: Function GWLS

[4.3] Examples
- One-dimensional normality
- Homogenization of a general data sample
- Comparisons with statistical approach
- Estimation of the membership function of a fuzzy-set
- Multidimensional normality
- Normality of a multidimensional model
- Personal diagnostic/reference bounds of the health state normality
- Normality bounds of contamination in rivers
- One-dimensional versus multidimensional normality

5. Conclusions of the BONUS project.

The aim of the project was simple, which is to rise the interest to the urgent problem of normality that relates to several branches of technology:

- Personalized medicine with reference values dependent on genetic factors.
- Environmental control with microcomputers checking each chimney emitting some gases and each tube discharging some liquids into rivers.
- Anti-fire systems.
- Warning and emergency systems of all types.

In all such applications the setting of bounds of normality should be based on robustly estimated values of homogeneous data samples established by using the gnostic methodology.

Each story has its end but the story of the BONUS project has still no end. It may be, that this book will bring a happy end.

12.5 Conclusions

The decision making on deviation from normal state represents a key problem in many fields of praxis. Independently if it is done automatically by technical means or relying on a person, the reliable decision making is a must because on its result not only material but also lethal dangers can depend. Statistical tests need reliable robust statistical models to be prepared a priori but the reality can appear to be different. Mathematical gnostics comes with several robust and reliable methods of checking the normality.

13

Applications of Global Distribution Functions

13.1 Global Distribution Function

There are two types of global probability distribution functions, the quantifying $QGDF$ and estimating $EGDF$ computed by using the gnostic kernels. The formulae are (10.1) through (10.8).

Soon after the first paper about mathematical gnostics was submitted [54], a short comment of R. H. Baran from the Naval Surface Weapon Center in White Oak, USA was presented to the editor informing on successful Monte Carlo tests of the gnostic distribution functions having the S-shape and the density without a sharp peaks. However, optimization algorithms existing already at that time enabled estimation of distribution of very different forms. An example is in Fig. 13.1.

There were several measuring points to register the impact of the Brno City (capital of Moravia) on the water contamination by the dangerous pollutant Triclosan:

1. Rivers upstream Brno,
2. Rivers downstream Brno,
3. Outlet of the Wastewater Plant,
4. Inlet of the Wastewater Plant.

All depicted distribution functions summarizing the results are far from the S-shape, especially the last one, the form of which could be called anti-S shape. This example shows like rich is the information provided by the distribution function:

- "Vertical": values of probability for different quantiles,
- "Horizontal": distances between individual distributions revealing
  - state of contamination before impact of the Brno City,
  - cleaning effect of the Wastewater Plant,
  - diluting of contamination by the river between the outlet of the Wastewater Plant and the second control place,
  - total contaminating effect of the Brno City and its periphery.

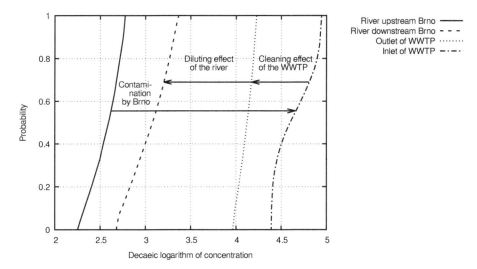

FIGURE 13.1
Application of EGDF to show contamination by the Brno City

- zero points on the lower axis: the minimum values at the measuring points,
- the points on the upper axis (probability 1): the maximum values at the measuring points.

The measurements were performed by laboratories of the Institute of Public Health, Ostrava under leadership of Dr. T. Ocelka and were supported by the grant *2-FUN*[1] of the European Union

The anti-S shape distributions are met frequently in real data studies relating to the fatigue cracks. There are other cases where this form of distribution appears as "normal" more than the "standard" S-form. An example is in Fig. 13.2.

The density of probability of the anti-S functions have the U-form showing that sharp limits of probability take place on both—the lower and the upper—sides of the data domain.

Another example is in Figure 13.3 showing the distributions of mean concentrations of led and cadmium in Silesian counties. Data were obtained from the Polish institute IETU Katowice.

The vertical straight lines denote the norms of the concentration. The distributions allowed to determine that concentration of cadmium was in about 30% and of lead about 12% exceeding the norm.

The availability of kernel estimates of global distribution allows a high degree of information capability of analysis to be reached. Consider the problem of comparison of 23 methods of measuring the toxicity of persistent pollutants

---

[1] https://cordis.europa.eu/project/rcn/81286_en.html

Global Distribution Function 137

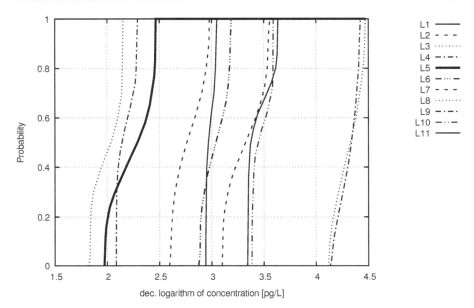

FIGURE 13.2
Concentrations of the pollutant Me-triclosan in rivers surrounding Brno City

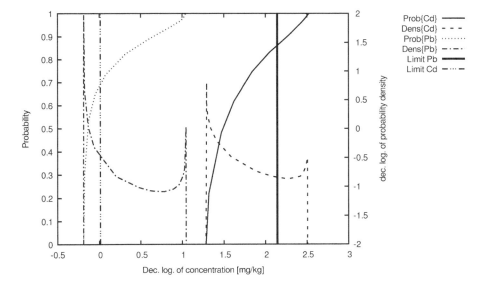

FIGURE 13.3
Distributions of Cd and Pb in Silesian counties

138   *Applications of Global Distribution Functions*

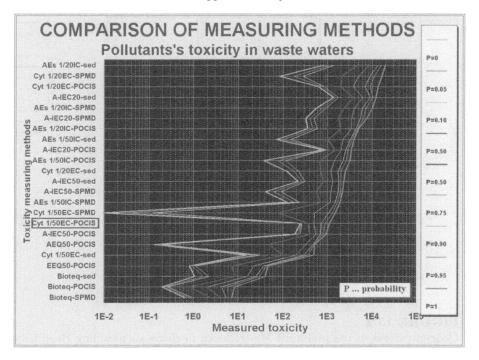

FIGURE 13.4
Results of 23 methods of measurements of toxicity

depicted in Fig. 13.4. The study was performed by the already cited team of the Institute of Public Health of Ostrava. Distribution functions *GEDF* of results of measurements were obtained and presented in the following figure.

The points of graphs correspond to the probabilities appearing on the right side of the figure. The method *Cyt 1/50EC-POCIS* has been emphasized because of its least range of the probability. However, this does not mean that it is the best method because many other methods could result with much larger values of toxicity.

This last example deserves a comment in respect to the high level of activity of the environmental research in Ostrava which is the "heart" of the Moravian and Silesian Region denoted *MSR*. This city along with the whole region became confirmed highly industrialized in the last two centurie. Its large concentration of the heavy steel and machinery enterprises using the rich resources of coal substantially helped the development of the whole state. However, the industrialization is also causing the degradation of the state of environment. This is reflected by the shortening of the lifetime of the people. This is shown in Fig. 13.5 where are the probability distributions of mortality of the people in *MSR* compared with that of the whole Czech Republic.

The intensive attraction of the environmental problems in this part of Central Europe has thus reasons of the fatal importance. About 80% of people

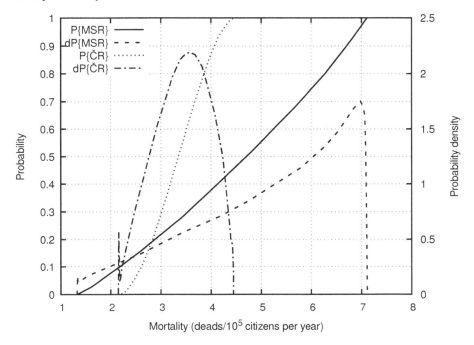

FIGURE 13.5
Mortality comparison of the MSR compared with the whole Czech Republic

living in Moravian-Silesian Region are under the danger of mortality exceeding that of people in the whole Czech Republic.

The geographic location of the Czech Republic in the Central Europe has an interesting consequence: it is the watershad divide: all its rivers have its sources in the country supplying the neighboring countries by water. The bad side is the responsibility of the Czech Republic for contamination of the "delivered" water. This is why the contamination of the water leaving the country is permanently under control. This example shows, how application of the probability distribution functions enables to concentrate a large amount of information in a figure. The results are related to the river called *Labe* in Czech and *die Elbe* in German.

13.2 Comparison of Global with Local Distribution

Availability of the global distribution function ($EGDF$) and the local one ($ELDF$) may lead to a question which should be used in case of a particular

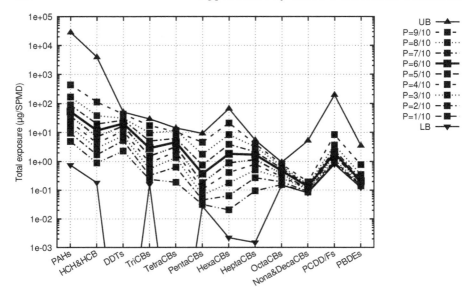

FIGURE 13.6
Contamination of the Labe river on Czech border

data sample. The answer may be general because it depends on the data. There may be different situations:

1. Clear homogeneous data sample, the application of the *EGDF* confirms the homogeneity.
2. Clear non-homogeneous data—use the *ELDF*.
3. Unknown data sample.

The application of the global distribution function is recommended in the first and third case, because the decision is dependent on the result of analysis. If the *EGDF* shows the non-homogeneity of the sample, the analysis is finished by *ELDF*.

13.3 Two Didactic Stories

Two stories relate to another industrial user of first gnostic software, the *TATRA* works who was famous for successes of his air-cooled heavy-duty tracks in international Dakar race. To start the academic-industrial cooperation, a seminary in the fabric was organized. A coworker of the quality control department prepared a data series for demonstration of the gnostic analyzer. To a great surprise of all, the public the probability density curve of their data has shown three maxima. But the quality inspector gave the compliment to the authors of the software for the presented result and explaining the cause of the maxima. The data were obtained as statistics of lifetimes of angular belts of the truck's drivers rotating the alternator. There were strict instructions for exchanging the belts: it was prescribed, that the alternator must be made free for the exchange and the tension of the belt after fixing the alternator back must be adjusted by a measuring device and must be within a tolerance. The data were obtained under three different operating conditions:

1. In the factory.
2. In authorized workshops.
3. On the street after the belt cracked.

The quantiles of the maxima corresponded to the place of exchange. The longest life-time was achieved with the primary installation under quality control. The shortest life-time resulted from the in a hurry exchange on a road. The experience of this application helped to decide on the broad applications of the gnostic software in the laboratory. At that time—before privatization— a development and research laboratory existed in the *TATRA* establishment where top specialists took care on the production quality and were cooperating with the designers of cars. They had experience with statistical methods and were glad to get new data analysis tools especially after the first success presented at the workshop. They introduced the gnostic analyzer into the daily use. But they contacted the author of the gnostic analyzer as there are strange results while comparing two technologies of a spring, the first and an alternative. "There is something wrong with your analyzer which shows the life-time of the first spring substantially longer that of the alternative. It is impossible because to increase the spring's life-time we added in production of the alternative a second life-time additional operation consisting in bombardment the spring's surface by bearing balls. The fatigue cracks are usually originating in surface cracks and the bombing helps to reintegrate the surface cracks." The advice was simple: check the way, how actually the second bombing was implemented. Such a verification really helped: it appeared that for the second bombing there were already no bearing balls on hand. Therefore cut steel wires were applied which instead of reducing the existed surface cracks produced new ones with its sharp edges.

There is an important lesson from these stories. The data never lie when they are "let them speak of themselves." However, the necessary condition for understanding what they "say" is knowing the true conditions under which they originated.

13.4 Conclusions

Applications of the global probability distribution functions enable several preferences before using the point estimates to be obtained. Their suitability for estimating the finite bounds of the data domain, their double kind of robustness and reliability in homogenization make them an ideal instrument for data analysis revealing not only the required information but also the unexpected information about the nature and state of the data sources.

14

Data Censoring

Although not stated explicitly, there has been a hidden assumption with respect to all the data sets still considered: they were made up of the exact (true) values and of the uncertainty, the probability density of which (the "kernel") was characterized by a function symmetrical with respect to the true value. Such an assumption is not justified with all data, because additional uncertainty exists caused by the effect called *data censoring*. It is important to note, that this type of censorship has nothing common with a subjective actions of an official against the human's rights. Data censoring considered here is connected with an objective additional lack of a priori knowledge about the actual data values. Let A_k be the k-th actually observed data value and $A_{k,u}$ its value, that would be observed without the effect of censoring. Four cases can be distinguished:

1. Uncensored data: $A_k \equiv A_{k,u}$,
2. left-censored data: $A_{k,u} \leq A_k$,
3. right-censored data: $A_k \leq A_{k,u}$,
4. interval data (censored from both sides): $A_{k,L} \leq A_k \leq A_{k,U}$,

where $A_{k,u}$, $A_{k,L}$ and $A_{k,U}$ are a priori established bounds. Examples of these bounds will be presented below. Taking the censoring into account starts with the uncensored case by showing how an estimate of a data item is interpreted with regard to the probability $p(A, A_0, S, AL, AU)$ of the true value A_0, where the AL and AU are bounds of data support.

14.1 Uncensored Data

Denote $\Theta(A) = dP/dA$ the probability density of $p(A, A_0, S, AL, AU)$ over the finite support (AL, AU) of an additive observed/measured data A, given true data value A_0 and scale parameter S.

Then the contribution of a datum A_k (the a priori weight of which is W_k) to the density of the empirical distribution function is given by the following statements:

$$(\forall A_{k,u})(AL < A_{k,u} < A_k)(\Theta(A_{k,u}) = 0) \qquad (14.1)$$

and
$$(\forall A_{k,u})(A_k < A_{k,u} < AU)(\Theta(A_{k,u}) = 0) \tag{14.2}$$
hold to satisfy the previously accepted condition
$$\int_{AL}^{AU} \Theta(A) dA = W_k. \tag{14.3}$$

These equations describe function Θ as an "impulse" function, integral of which adds a step (W_k) to the weighed distribution function in the point A_k. Such a step is caused by an uncensored data item, for which $A_k \equiv A_{k,u}$.

14.2 Left-Censored Data

Despite the constant effort and improvements in analytical chemistry, the quantification of low-level contamination levels in environmental materials like water, air, soil and/or food, and in human materials monitored for biomonitoring issues may remain difficult for many substances. Thus, datasets used in environmental and/or human risk assessments are commonly characterized by a significant fraction of values below the so-called Limit of Detection (LOD) It is also known that environmental and/or health effects can be induced by low-level chronic exposure of single substances or substances in combination (e.g. dioxins). In such contexts, it is then essential to affect realistic (even actually unknown) values to non-detects (i.e. data below LOD, called also Type-I censored data. Several approaches were proposed so far to handle this issue. The most common (and easy-to-use) approach is based on substitutes: non-detects are discarded from the dataset, or substituted by constant values like zero, LOD itself, $LOD/2$ or $LOD/\sqrt{2}$. For example, US-EPA provided a critical guidance of the reliable use of such methods in its Guidance for Data Quality Assessment (EPA, 2000) and suggests that substitute methods could be used until a 15% fraction of non-detects in a given dataset. It is however obvious that such methods lead to over- or underestimation of the mean, and under- or overestimation of the standard deviation (for zero and LOD substitutes respectively), or they generate bias depending on the skewed nature of the dataset. Other methods, called "distributional" methods were also developed, but they require to define assumptions on the underlying data distribution (e.g. Normal, Log-normal, Gamma or Gumbel distributions). Parameters of the assumed underlying distribution (e.g. mean and standard deviation for a log-normal distribution) are derived from the maximum-likelihood estimation. Distributional methods can also be extended to reconstruct a complete dataset affecting potential data to non-detects (distribution-based imputation). In the absence of knowledge about the underlying distribution, the use of such methods remains however questionable. Other methods were proposed in some specific applications: for example, analogies between substances can be assumed to reconstruct data for non-detectable substances from detected ones. This is the approach proposed for dioxin and furan congeners, but this "congener ratio approach" cannot be extended to other substances and other regions than those specifically observed.

Left-censored data can also be encountered in other fields than in analytical chemistry. Examples:

LIFO inventory valuation in periods of rising prices: Under the LIFO (last-in, first-out) method, the costs of the last goods purchased are charged against revenues as the cost of the goods sold, while the inventory account is based on the costs of the oldest goods acquired. When prices rise, the inventory is thus undervalued.

The cost of acquisition of an item of inventory can sometimes represent the upper bound of possible prices of an item in inventory such as goods, which are out of fashion or technologically below the current state of the art.

Expert's underestimation e. g. "not worth more than A_k."

Insufficient sensitivity: Imagine an automatic measuring system installed to check the concentration of dangerous gases. Some results are below the detection threshold of the sensor, but they cannot be ignored, because they represent the most desirable conditions of the monitored system. Such measurements can be therefore treated as left-censored data.

The selling price of goods is in most cases higher than the upper estimate of the goods true value. If its price is T_k than one can interpret it as "the actual value is anywhere between zero and T_k."

In applications of the $EGDF$, the contribution ($\Delta WEDF_k$) of left-censored data A_k to the $WEDF$ (Weighted Estimating Distribution Function) can be evaluated in the following way:

$$(\forall A \leq A_k)\left(\Delta WEDF_k = W_k \frac{A - AL}{A_k - AL}\right) , \qquad (14.4)$$

$$(\forall A > A_k)(\Delta WEDF_k = W_k) . \qquad (14.5)$$

It is interesting to compare several methods of treating the left-censored data in application to real data. In (14.1) the distribution functions are obtained by the methods cited above in comparison corresponding to the distribution function $EGDF$ denoted as "Censored." The data are concentrations of the pollutant $TCDD$.

As shown in the figure 14.1, the errors of non-adequate methods of censoring may be significant. This can be documented by data from a different field.

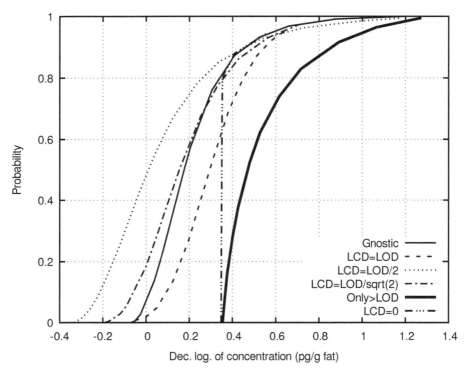

FIGURE 14.1
Comparison of methods of treating the left-censored data; LCD – left-censored data, LOD – limit of detection

However, there exists a statistical method of Kaplan-Meier whose results are comparable with those of gnostic method. This method is generally accepted in statistics and applied by famous clinics. The data for this example stem from regular monitoring of contamination of Czech rivers, particularly by *GammaHCH*.

This comparison shows close coincidence of the continuous gnostic curve with the lower and upper step functions of the K-M results. It means that the gnostic method works at least as the famous statistical approach. The important difference is that the gnostic approach enables the fatal finite point to be numerically predicted.

14.3 Right-Censored Data

It is possible in practice, that not only the number A_k itself is given as a member of the data set to be treated, but that additional information about

Right-Censored Data

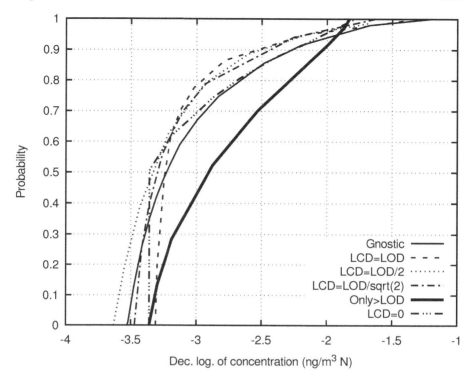

FIGURE 14.2
Methods of treatment the left-censored data applied to exhalations of furnaces; LCD – left-censored data, LOD – limit of detection

it is available from a knowledge of the measuring process or from the nature of the particular datum. Such a situation can be defined as:

- Condition (14.1) holds.

- Instead of (14.2), statement

$$(\forall A_{k,u})(A_k \leq A_{k,u} < AU)\left(\Theta(A_{k,u}) = \frac{W_k}{AU - A_k}\right) \quad (14.6)$$

characterizes the probability density of the event, which was quantified by A_k. Examples of right-censored data include:

Survival data: Positive data T_k are the life-times of a group of objects. A "life-time" is the span from the object's "birth" T_0 until its "death." An uncensored lifetime is defined completely by the number T_k, which is fixed by the already known end of the object's life. However, in all the rest of the cases T_k is used to describe only the end of the observation period, not the

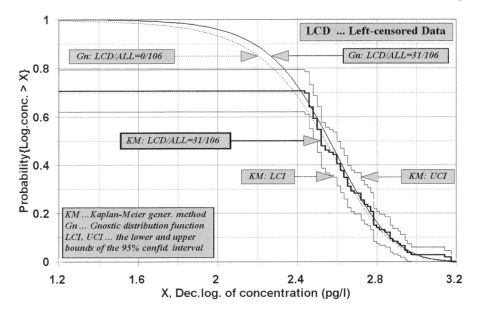

FIGURE 14.3
Comparison of the Kaplan-Meier method with gnostic approach

end of life. The object 'lives on': it is only known, that it "survived" time T_k, but there is no evidence of its "death." These data are to be read as "the life-time of the k-th object is **at least** T_k."

Prudence in accounting: The principle of conservatism in accounting means being cautious or prudent and making sure, that net assets and net income are not overstated. The entries in the corresponding financial statements are therefore rather right-censored than uncensored.

Measurements off the scale: A measured variable exceeded the maximum value of the scale of the measuring system.

Tax expense: In the real world people tend to estimate their tax obligations on the low side. This means, that tax paid (T_k) can be interpreted as "actual obligation was **at least** T_k."

Gnostic models of continuous distribution functions are designed to provide the best goodness-of-fit of the set of points of the function *wedf*. To take into account the censoring, the *wedf*'s points $WEDF$ need to be revised. This is done in the case of right-censoring by the formulae

$$(\forall A < A_k)(\Delta WEDF_k = 0) \;, \tag{14.7}$$

$$(\forall A \geq A_k)\left(\Delta WEDF_k = W_k \frac{A - A_k}{AU - A_k}\right) \tag{14.8}$$

Interval Data 149

where AU is the upper bound of the data support, and where $\Delta WEDF_k$ denotes the data item's A_k contribution to the function $WEDF$.

14.4 Interval Data

The judgment of members of an expert board can just as easily either over- or underestimate the value of an object. Each point would normally represent an uncensored datum, however, estimates could be in the form of "not less than A_k, but not more than A_m." In other words, the data item could be of interval nature, censored from both sides. Such data are not as rare as it would seem. Consider some nontrivial examples:

Product's quality: The distribution function of a parameter Q characterizing the quality of comparable products from different producers is to be calculated. There are two types of data possible:

- Measured values of Q (the uncensored data).
- Specific results of measurements of Q, values of which are not provided, but that are only counted—due to a reliable automatic quality control system—to surely fall within the interval $[Q_k, Q_m]$. Such data would then be accounted for as interval data.

Market prices are based on estimates of the true current value of goods. Assume, that both parties to a deal are well informed about the market situation. It is natural to expect, that the asking price (set by the seller) will be more than his estimate of the true value, while the bidder (the buyer) will try to establish the price under his estimated true value. The bid price can thus be viewed as the lower and the asked price as the upper bound of the interval of acceptable prices.

Multichannel analyzers There are instruments in kernel research, registering not the energy of particles, but only its belonging to an interval of energies.

Three relations account for an interval data item spread from A_k to A_m:

$$(\forall A < A_k)(\Delta WEDF_k = 0) \,, \tag{14.9}$$

$$(\forall A)(A_k \leq A \leq A_m)\left(\Delta WEDF_k = W_k \frac{A - A_k}{A_m - A_k}\right) \,, \tag{14.10}$$

$$(\forall A > A_m)(\Delta WEDF_k = W_k) \,. \tag{14.11}$$

The function $GNDF$ enables the censored data's true values to be estimated by setting its argument's *Censoring* and *Bcensup*.

14.5 On an Unknown Limit of Detection

Preparation of the application of the distribution functions involves a requirement to fill in the data columns *Censoring* and *Bcensup* with symbols defining the way of data censoring. The column Censoring distinguishes four groups of data: the left- and right-censored data, the uncensored and interval data. The decision relates to the input data value D_i and to the value of *Limit of Detection* (*LoD*). (We shall distinguish the lower limit of detection *LLoD* and its upper value *ULoD*.) The data item is declared as left-censored if $D_i < LLoD$, as right-censored if $D_i > ULoD$ and otherwise as uncensored or interval data item. The interval item has in the column *Bcensup* the upper limit of its interval value while as its lower limit is the data value D_i substituted. This value is also used in the case of the uncensored data item as the quantile to locate the corresponding kernel contribution to distribution function. Locating of a kernel in the case of a censored data item is more complicated. Let us consider the left-censored items. They are ordered by the values D_i and the interval $D_{L,i}$, $D_{U,i}$ between the smallest ($D_{L,i}$) and largest ($D_{U,i}$ values is divided by the number NL where NL is the number of left-censored data). The "measured values" may be equal, but their number is what is important. The new values $D_{L,1} \ldots D_{U,NL}$ located of the knots of the uniform distribution over the interval between the *LLoD* and the least uncensored value are then applied as quantiles for placement of the lower rest of kernels' quantiles. The upper rest of kernels is located analogously.

However, this approach assumes that the values of the *LLoD* are known. Such an assumption is in order in many practical cases when the measuring person is able to guess this value by experience. But there is a chance to obtain the *LLoD* objectively by using the data due to the high robustness of the distribution function *EGDF* iteratively:

1. To apply the first guess of *LLoD* in denoting the values of the column *Censoring*.

2. To run the function *EGDF*.

3. If the function appears to be non-homogeneous, adjust the value of *LLoD* by adjusting the values in *Censoring* and retry the point 1.

Some few iterations of this process should suffice.

However, the estimation of the *LLoD* is a much more difficult task then that of *ULoD* which is officially declared as the *finite* maximum of scale. If the uncertainty u evaluates the measurement's error, then the relative error of measuring value of x which is u/x rises to infinity with x closing to zero.

14.6 Examples of Surviving

The degenerative changes in various tissues/organs have been attributed to derangement of stem cell functions causing regenerative tragedy. Bone marrow stromal cells (*BMSCs*) are considered the ideal candidates for regenerative approaches owing to their beneficial effects in numerous clinical applications. There is a rich experience in applications of the *BMSC*-methods, especially in healing deep burns and also in healing the diabetic patients in Faculty Hospital in Ostrava (Czech republic). Study [89] has been analyzing the data on diabetic patients, especially in connection with amputation of a leg. There was a control group of 56 such patients subjected to traditional mediating and 50 patients healing by application of *BMSC* (Bone Marrow Stromal Cells), both groups sub-divided into two sub-groups of amputated and non-amputated patients. The lifetimes of both groups was not perfectly comparative because the patients of the control group were under the care of the hospital for a longer time. The lifetimes were predicted by means of the gnostic distribution functions. The first unexpected result was that of comparison of the lifetime predictions of amputated and non-amputated patients treated by means of the *BMSC* method shown in Fig. 14.4.

There is a critical question for a heavy diabetic patient when his doctor has to take a decision about cut his leg: to be or not to be. The results shown support this interpretation of the decision. The positive answer really

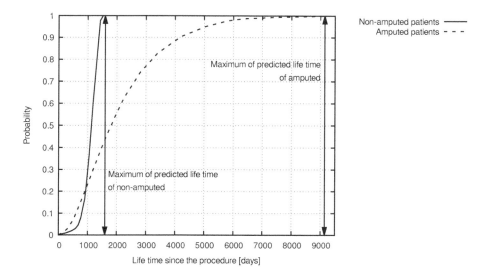

FIGURE 14.4
Lifetimes of amputated and non-amputated patients treated by *BMSC* methods.

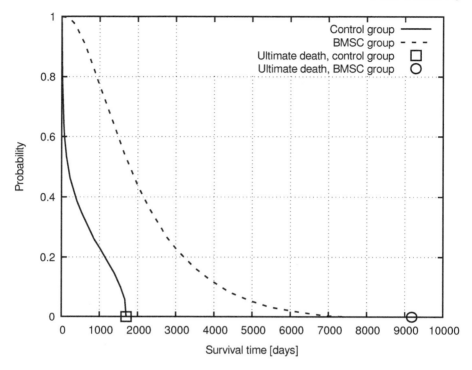

FIGURE 14.5
Comparison of predicted lifetimes of *BMSC* and traditional treatment

means "to be" for a substantially longer lifetime. The next figure documents the advantage of *BMSC* methods over the traditional ones at least for amputated patients: the lifetime of amputated patients treated with *BMSC* was substantially exceeding that of the traditionally medicated.

This result documents the prevalence of the *BMSC* methods over the traditional ones.

Some information connected with the survival problem can be obtained directly by the probability distribution function thanks to its capability to estimate the bounds of data support. One example can be shown from the environmental control of contamination of rivers by the biotic method: fishes caught are analyzed to determine arsenic accumulated in their organism during their life. The probability distribution reveal the maximum amount found.

The zero probability of exceeding this value has a clear interpretation: no fish was able to survive it because this value is the lethal dose of arsenic consumed by fish.

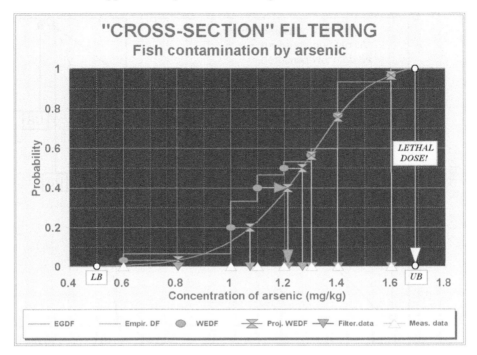

FIGURE 14.6
Distribution of arsenic in fish

14.7 Non-Standard Application of Data Censoring

14.7.1 Data and Psychology

The primary goal of measuring is to find the true value of a quantity. It could seem that such an activity is an entirely objective matter. However, such a conclusion would be false because there are data representing a good message and bad message and the data receiver is a human. It means that there also are psychological aspects of data:

Contents: "Good" data inform on positive aspects of life, on success, wealth, gain, victory, property, health, prosperity, freedom, peace. Unlike this, there are "bad" data that inform on dangers, losses, fails, diseases, disasters. The former are welcome, the latter not.

Plausibility: This aspect depends on the status of data source as well as on the data value.

Impact: Data can attract interests of the receiver and/or cause his decision or actions.

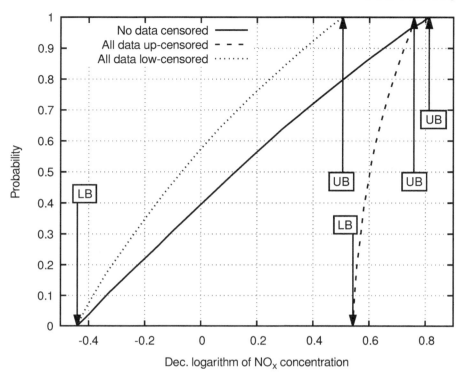

FIGURE 14.7
Three interpretations of the exhaust of an ethanol engine

In spite of the worth of knowing the true value of a variable, its knowledge is accepted individually by individual receivers. The technique of data censoring brings a chance to exercise a personal interest in interpretation of objectively determined true data and to evaluate the probable bounds of the alternative. The idea is to take the actually measured data from three points of view: as they are, as the upper estimate of actually left-censored data and as the lower estimate of actually right-censored data. The first approach can be called realistic while the second and third is pessimistic or optimistic based upon the dependence on the character of the message born by the data. Such evaluation represents a decline of objectivity but offers additional gains.

14.7.2 Three Aspects of Data Interpretation

The data of testing an ethanol drive can be found in the database of the R-project.[1] The exhaust of NOx was measured and the results presented in Fig. 14.7 as a nearly linear probability distribution function with its bounds LB and UB.

---
[1] www.r-project.org

Non-Standard Application of Data Censoring

To use the measured data is the realistic interpretation. However, the company preparing the engine for the market takes the measured values as the "at most" reached. To this "optimistic" interpretation the company runs the evaluation as of the left-censored data shifting the distribution function to the left. But there might be an opposite point of view: the officer of the EPA (the Environmental Protection Agency of USA) remembering the scandal with exhausts of German cars will apply a pessimistic point of view and interpret the measured data for right-censored ones, as the "at least" reached. The probability distribution is shifted to the right.

The relation of firms and individuals to tax paying has been already discussed. Actually paid taxes can be therefore taken as the right-censored data. The task for a tax office is not only take care for proper tax paying. The technique of censored data allows them to get the estimate of the maximum possible tax yield by running the actually paid data as right-censored. But there is another point of view, that of tax payer: what would be if their actually paid taxes would be taken as left-censored. Effects of both subjective points of view can be evaluated by means of Fig. 14.8. The data for this

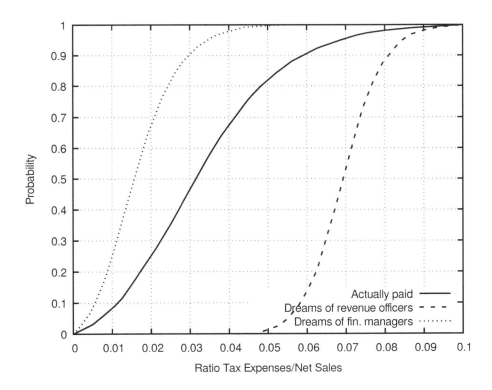

FIGURE 14.8
Three interpretations of the tax paid

example stem from the financial statement data of the chemical industry quoted on the market of USA in 1998.

It might be interesting to note, that the horizontal differences between the distribution functions showing the effect of censoring increase with rising tax paid at the left side while the opposite is true for the other side. It is a lesson for the tax office: to rise the tax paid, take care for all tax payers including the small and the big ones. The effect of censoring is significant, the medians are roughly in proportions 0.5, 1.0, 2.0.

The opposite interests of buyers and sellers were already also discussed. The actual price of a good is result of balancing the interests of both sides of the transaction. To illustrate this problem, data from the financial market can be used from the time, when the Deutsche Mark existed and was traded for US dollars.

The market was running as the international in-line system operating between special national banks. Two variables were followed as actual market information, bid and ask of the exchange ratio. Two "outer" distribution functions in Fig. 14.9 are showing bid and ask. The *BID* curve represents the ratio which is the buyer decided to accept, but it is considered as the right-censored

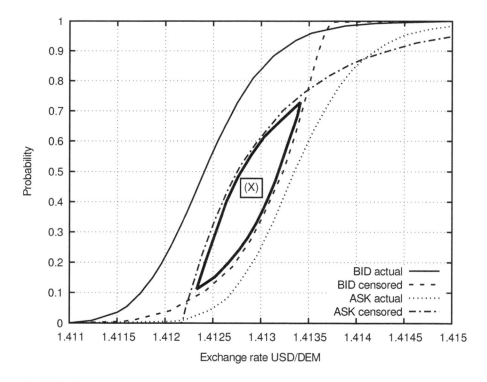

FIGURE 14.9
Three interpretations of the trading; (X) – area of acceptable compromise

value. The *ASK* curve shows the asked value of the ratio which is taken as left-censored. The inner probability curve shows the uncensored values of BID and ASK. The curves delimit the range of reasonable ratios for transaction. Such curves can thus be used as recommendation on the reasonable bounds of prices. Another application of this type of data treatment can be in the estimation of the bounds of risks based on the distribution function of actual risks. Such an estimate could be of help in establishing the rational prices of insurance.

14.8 Conclusions

Censored data are non-completely measurable data. There are three categories of such data, the left- and right-censored and interval data. The left-censored data are data measured below the lower bound of the measuring range and the right-censored data are data measurable only below the upper bound of the measuring range. The interval data are data values of which are identified only as being between two limits. There exist typical applications of all three data categories. It is an important feature of gnostic probability distribution functions that they are able to estimate the values of all censored data met on practice. Some curious applications of censored data are enabled as well by introducing the factor of subjective interpretation when real uncensored data are artificially interpreted as left- or right-censored ones. This approach enables to see the data in a pessimistic, realistic or optimistic way and estimate even the bounds of reasonable trading prices or bounds of prices of rational insurance.

15
Gnostic Thermodynamic Analysis of Data Uncertainty

The analysis of the Ideal Gnostic Cycle showed three thermodynamic interpretations of the data uncertainty:

Q The thermodynamic equivalent of quantifying rise of entropy measured by the change of quantifying uncertainty $f_J - 1$,

E The thermodynamic equivalent of estimating decrease of entropy measured by the change of estimating entropy $1 - f_I$,

R The residual entropy of the Ideal Gnostic Cycle measured by the difference $f_J - f_I$.

All three thermodynamic equivalents of uncertainty can be used as criterion functions for optimization of operations of data treatment to minimize the residual entropy of the IGC by

Q ... minimization of f_J,

E ... maximization of f_I,

R ... minimization of $f_J - f_I$.

There are several gnostic functions implementing these ideas. However, the most prospective method is using the thermodynamic response to an additional data item taking variable values on a large interval: the gnostic data calibration.

15.1 Gnostic Data Calibration
15.1.1 Real Data for Examples

The conclusiveness of examples is dependent on the data applied. Therefore the data selected for the examples were taken of the official publicly available database of the NIST (The National Institute of Standards and Technology

USA). According to the https://www.nist.gov/srd, *NIST produces the Nation's Standard Reference Data (SRD). These data are assessed by experts and are trustworthy such that people can use the data with confidence and base significant decisions on the data. NIST provides 49 free SRD databases and 41 fee-based SRD databases. SRD must be compliant with rigorous critical evaluation criteria.*

Two real datasets were selected for the examples called here for simplicity namely NIST12 [38] and NIST37 [39]. Their short characterization follows:

NIST12 Data set CAS RN 110-56-5 (Normal boiling temperature of 1,4-dichlorobutane). 12 data items, NIST summary is 410 ± 80. Statistical evaluation: (mean ± STD): 410.28 ± 44.68. The data sample has been shown to be non-homogeneous by the global distribution function. The outliers were identified as Nos. 5 and 9.

NIST12H This is the homogeneous sub-sample of the NIST12 consisting of 10 data (Nos. 5 and 9 excluded). Statistical evaluation: (mean ± STD): 417.51 ± 38.57.

NIST37 Normal boiling temperature of chloroform. 37 data items, NIST summary is 334.3 ± 0.2. Statistical evaluation: (mean ± STD): 334.31±0.11.

It may be interesting to have a look at data NIST12 before (Fig. 15.1) and after homogenization.

Non-homogeneous NIST12   Homogeneous NIST12H

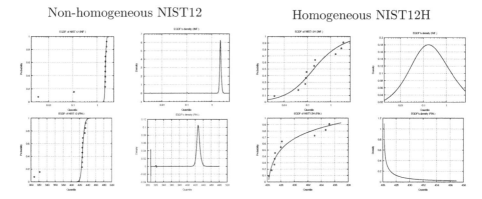

FIGURE 15.1
Data NIST12 before and after homogenization

The second maximum cannot be clearly distinguished in the left graph of the *DistribNIST12*. It is necessary to look at the numeric values of the probability density over the infinite data support at the right side of the graph. There is a clear maximum 6.953193e-01 at point number 299.

The difference between the estimation error (80) and statistical standard error (44.68) could be explained by the form of the distribution function which

TABLE 15.1
Course of the probability density over the infinite data support of the *Distrib-NIST12* before and after homogenization

Data No.	Value	Density
288	423.1239	5.605893e-01
289	423.5537	5.812077e-01
290	423.9753	6.007449e-01
291	424.3890	6.189767e-01
292	424.7947	6.356842e-01
293	425.1926	6.506600e-01
294	425.5826	6.637137e-01
295	425.9650	6.746772e-01
296	426.3397	6.834102e-01
297	426.7069	6.898039e-01
298	427.0666	6.937855e-01
299	427.4190	6.953193e-01
300	427.7641	6.944087e-01
301	428.1021	6.910954e-01
302	428.4330	6.854582e-01
303	428.7569	6.776095e-01
304	429.0739	6.676920e-01
305	429.3842	6.558735e-01
306	429.6878	6.423417e-01
307	429.9848	6.272984e-01
308	430.2753	6.109535e-01
309	430.5594	5.935196e-01

is far from the Gaussian. However, the *NIST*'s error value is still much less than the range of NIST12 data (127.05). The question of an estimation error stays to be open. As an approximation of the estimation error the standard error of the homogenized sub-sample of the NIST12H data (38.57) could not serve, because the gnostic distribution function has a limited data support (estimated as [426.2059, 437.3518]) the length of which is not far from the range of homogenized data NIST12H (8.95). It is obvious, that the look of the data by means of the gnostic homogeneous distribution functions provides more certain information than the statistical look.

However, a look at this homogenous distribution function (at small quadrates showing the positions of Kolmogorov-Smirnow points on the empirical distribution function) indicates the necessity to go deeper into the structure of the homogeneous sub-sample. This is done in the next chapter.

Data set NIST37 is already at first sight much more precise than the NIST12. However, its situation is actually even more complicated. Its global distribution function is in Fig. 15.2.

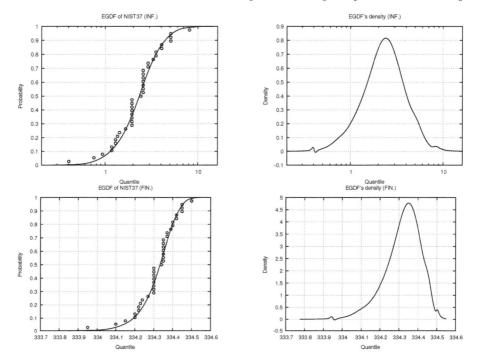

FIGURE 15.2
NIST37

A careful sight on the edges of the probability density function on both the finite and infinite data support reveals additional maxima: the data sample is not homogenous. It is worth of applying the local distribution function with the scale parameter $S = 0.2$ (Fig. 15.3).

We can see nine peaks on the figure which correspond to the structure of the data sample composed of only nine "solo" values while five times are the values doubled. Three times are values repeated once, while values 334.30 appear eight times and 334.35 seven times. Which value should be chosen as the standard value? The aim to homogenize the sample by means of leaving out the "outliers" leads to the homogeneous sub-sample consisting of the value 334.30 repeated eight times. However, seven values of 334.35 also forms a homogeneous sub-sample. The $EGDF$ of concatenation of both homogeneous sub-samples denotes all the seven values of 334.35 as upper outliers with respect to the sub-sample of eight values 334.30. The results of this analysis indicate that the "reference value" should be 334.30. It is interesting that statistical evaluation of concatenated both homogeneous sub-samples of the form (mean ± STD) is (334.3132 ± 0.0258). This result "supports" the decision 334.30 while median value 334.34 is closer to 334.35.

Data Calibration 163

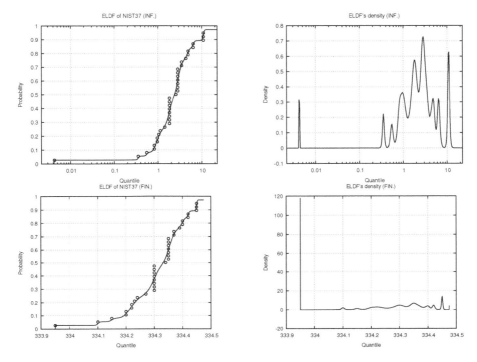

FIGURE 15.3
The local distribution of data NIST37

15.2 Data Calibration

The idea of data calibration is motivated by the successful application of an additional variable in interval analysis and by the aim to make use of thermodynamical properties of the data uncertainty. However, the interval analysis needs the local distribution function of a homogeneous data sample while data calibration is applied directly to an one-dimensional data sample without a priori guarantee of its homogeneity. The object of our attention is the behavior of the residual entropy RE of the Ideal Gnostic Cycle, its reaction to changes of an additional data item along the whole range of the data support of the sample to be analyzed. We can observe not only the characteristics f_I and f_J and their difference, the residual entropy RE, but also the RE's rate of change and its acceleration to identify the extreme points, zeroes and inflexion points of the observed quantities. To perform the operation of differentiating the time series the technique of numerical operators was applied. It is worth mentioning this useful numerical approach.

15.2.1 LS-Optimal Numerical Operators

There was time when to perform numerical calculations you needed monstrous mechanism which really deserved to be called mathematical machine. It was difficult to program it, because each individual operation was necessary to place on a performed card as a series of holes. Ideas capable to spare number of operations were very welcome at that time. But the necessity to minimize the number of operations endured even till the start of programmed transistor calculators capable to store a limited number of commands. It was a real art to write up a programme not exceeding, say, one hundred of commands. The idea of numerical operators was known but mathematicians did not see reasons for increasing the length of operators over the really necessary number. But such a reason existed, it was the necessity of improvement the operation quality when working with uncertain data. It has been shown in[1] that the most required operations to be applied to time series by the Least Squares method are linear with respect to the treated data. Moreover, in many cases the observations are distributed uniformly. This category includes such operations as estimation of coefficients of approximating functions (e. g. polynomials), of their derivatives and integrals, of interpolating values and predictions. The increase of the number of nods helps to minimize the impact of uncertainty. This allows the calculations to be split into two phases:

1. Computing the matrices solving the Least-Squares task.
2. Repeated applications of the resulting numerical operators to the data in the form of the scalar product.

The first phase is passed only once. The second phase is performed several hundred times for each type of operation in the case of for e. g. calibrating analysis. As seen in figures of calibration, the number of nods equal to 500 is sufficient for creating smooth functions by the application of numerical operators to the residual entropy to obtain its interpolated value, the first and second derivatives.

15.3 Calibration of the NIST12 Data

The first call of calibrating operation in application to the NIST12 data set is illustrated in Fig. 15.4.

The figure has four fields denoted as matrix elements. The field (1, 1) shows the means of functions f_I and f_J in dependence on values of the additional

---

[1] Kovanic P.: The Identification of Operators by Means of Static Programming (in English), IFAC Symposium "The Problems of Identification in Automatic Control Systems," Prague (1967), 19 pp. and in Kovanic P.: Optimum Digital Operators (in English), IFIP World Congress '68 (1968), Edinburgh

Calibration of the NIST12 Data

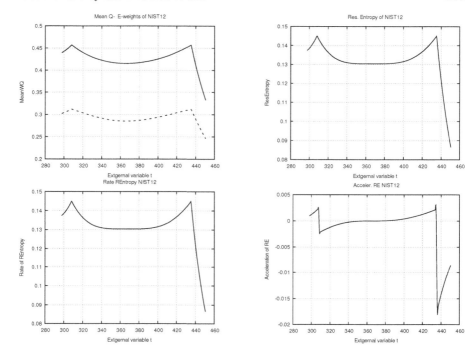

FIGURE 15.4
The first step of calibrating the data NIST12

(extending, "external") variable. Their difference RE (residual entropy) is in the field (top, right), the rate of change of the RE is in (down, left) and its acceleration in (down, right). The small circles on the x-axis denote the location of data. The black circles in curves show the extreme points of the curves and the triangles depict the points of inflection. The input data are decomposed into the *kernel* matrix and three sub-samples. The kernel matrix called NIST12K has the form of a Tab. 15.2.

TABLE 15.2
The kernel matrix

Row name	Location	Res. entropy
LCB	310.0746	0.1432496
CRE	430.4123	0.1412201
RCB	431.7214	0.1427563

The symbols LCB and RCB denote the rows of the left and right calibration bound (which can sometimes exchange the priority) and CRE denotes the calibration residual entropy. The data satisfying the conditions of the kernel form the central data matrix Fig. 15.3 called kernel NIST12K.

TABLE 15.3
The kernel data matrix

Id	NIST12K
5	322.00
8	426.25
12	426.65
10	427.05
3	427.10
6	427.15
11	428.00
4	428.15

Column Id conserves the original data names.

The lower sub-sample consists of data values which are below the lesser of LCB and RCB. The upper sub-sample is formed by data with values exceeding the LCB and RCB. This time, the lower sub-sample consists of only one data item NIST12L (9 308.15) while the upper is a matrix Tab. 15.4.

TABLE 15.4
The upper data matrix

Id	NIST12R
2	433.00
7	434.65
1	435.20

The process of data calibration continues with step-wise triplication of the central (kernel) data matrices till the moment, when the kernel data matrix is empty or consisting of the same data values. In this way, a sequence of kernel matrices is obtained of the form of Tab. 15.5:

Following remarks to these results are noteworthy:

1. The data were obviously measured with the resolution power of ± 0.05. The NIST standard value was given as 410 ± 80 while the statistics gave the result in the form of mean $\pm$ STD 410.28 ± 44.68. The estimation error was exceeding the measuring tolerance many times. Unlike this, the final tolerance interval obtained by means of data calibrating (difference $RCB - LCB$) 0.0959 could be taken as twice the error of estimation of $(RCB + LCB)$ as comparable with the sensitivity of measurement ± 0.05.

TABLE 15.5
The series of kernel matrices of NIST12

Sub-sample	LCB	CRE	RCB	Stat.mean
NIST12				410.2792
NIST12K	310.0746	430.4123	431.7214	414.0437
NIST12KK	323.6590	424.0713	425.3235	434.2833
NIST12KKK	426.2842	426.7143	428.1057	427.1929
NIST12KKKK	426.6743	427.9685	427.9686	427.1900
NIST12KKKKK	427.0518	427.1476	427.1477	427.1000

2. The statistical mean was gradually closing to the calibrated value during the process of calibration.

3. The sequences of the bounds LCB and RCB can be used for establishing the *quality class* of the quantity represented by the data value. This quantity can be a qualitative parameter of a product, result of a laboratory in a comparison of laboratories, evaluation of a sport competition, quantification of a danger for the health and others.

4. The calibration approach—unlike the interval analysis requiring the homogeneity—can be applied to arbitrary data sample, both homogeneous and non-homogeneous.

15.4 Calibration of the NIST37 Data

The NIST37 data set is much more precise than NIST12, the least non-zero difference between its data is 0.01, which represents precision class of 1/334%. However, the sample is non-homogeneous. Its extreme values are 333.95 and 334.50, but there are some local clusters (as we have seen in the local distribution function Fig. 15.3) located so far from each other that they cannot be "packed" by a common global distribution curve. The more is interesting the calibration analysis of this data sample.

There was eight steps of calibration necessary till the kernel data appeared to have equal values. The sequence of the kernel matrices had the following form:
The bounds LCB and RCB can be used to data classification Tab. 15.7.

The identifiers (Ids) of members of the sub-samples are the row names of the data in the original data sequence. The data classes correspond to steps of calibrating triplication when the data items are separated from the kernel data as the lower or upper ones.

TABLE 15.6
The series of kernel matrices of NIST37

Kernel sub-sample	LCB	CRE	RCB	Stat.mean
NIST37				334.3132
NIST37K	333.9599	334.4871	334.4872	334.3183
NIST37KK	334.1063	334.4418	334.4419	334.3126
NIST37KKKK	334.1549	334.4136	334.4137	334.3107
NIST37KKKKK	334.2036	334.3953	334.3953	334.3125
NIST37KKKKKK	334.2231	334.3860	334.3860	334.3176
NIST37KKKKKKK	334.2325	334.3667	334.3667	334.3167
NIST37KKKKKKKK	334.2420	334.3474	334.3474	334.3010
NIST37KKKKKKKKK	334.2713	334.3383	334.3384	334.3000

TABLE 15.7
Classification of the NIST37 data

Ids of members	Value	Class
11	333.95	-9
4	334.10	-8
34	334.15	-7
20,30	334.20	-6
22,31	334.22	-5
26	334.23	-4
10	334.24	-3
23	334.27	-2
1,8,17,18,24,25,27,33	334.30	1
6	334.34	+2
5,9,14,15,16,32,36	334.35	+3
2,29	334.37	+4
7	334.39	+5
28,35	334.40	+6
12,13	334.42	+7
3,19,37	334.45	+8
21	334.50	+9

The final difference between the bounds LCB and RCB is now 0.0671. This enables to represent the "calibrating mean" of the NIST37 data set as 334.305 ± 0.034.

The whole data calibration process consisting of all triplicating operations is depicted in Fig. 15.5.

Each triplicating step results in three matrices, L—the lower, K—the kernel matrix and U—the upper matrix. Information obtained from the

Conclusions

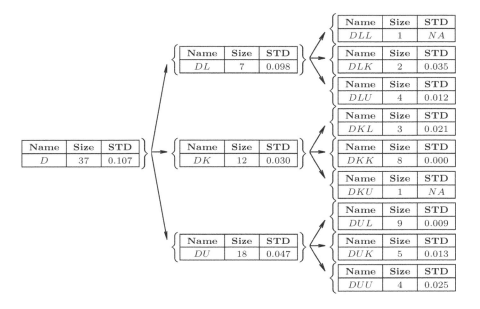

FIGURE 15.5
The whole data calibration process of the NIST37 data

non-central matrices might be also useful in analysis of the data subjected to calibrating triplication.

This application of the word triplication should be distinguished from the results of functions Gmean and Contribs.

15.5 Conclusions

The method of data calibration consists of analysis of reactions of residual entropy changes caused by data variability to the values of an additional variable changing over a broad interval. The analysis of extreme points of the curve of residual entropy provided by the numerical operators enables to triplicate the analyzed sample to get the lower sub-sample, the kernel and upper sub-samples separated by bounds corresponding to local extremes of the residual entropy. Repeated triplication of the kernel data enables not only the final "central" kernel value to be obtained but also series of bounds useful for detailed classification of sub-samples of the data set. The resolution power of the calibrating classification is comparable with that of data measurements.

16
Robust Estimation of a Constant

16.1 Gnostic Data Aggregation Principle Used in Estimation

The general way of making use of the gnostic data aggregation includes following steps:

1. Test if the data to be treated are additive or multiplicative, which of metrics 'Q' (quantifying), 'E' (estimating) or their combination 'R' is the most suitable for the given data and if the scale parameter should be constant or data-dependent.

2. Transformation of the additive to multiplicative data,

3. Solving the optimization problems:

$$f_J = \min_{Z_0} f_J(Z/Z_0) \tag{16.1}$$

where Z_0 is the true value to be estimated and denoted GMQ, while f_J is the quantifying uncertainty (2.9), and

$$f_I = \max_{Z_0} f_I(Z/Z_0) \tag{16.2}$$

where Z_0 is the true data value to be estimated and denoted GME, while f_I is the estimating certainty (2.15).

$$f_I - f_J = \min_{Z_0}(f_I - f_J)(Z/Z_0) \tag{16.3}$$

to minimize the residual entropy and to obtain the estimate GMR.

4. Transformation of the mean values back to the original additive or multiplicative data support.

5. Estimation of the scale parameter or scale parameters.

6. Evaluation of statistical mean and its standard error to write the result in the form of *mean* ± *mad* where *mad* is the mean absolute deviation from the median.

DOI: 10.1201/9780429441196-16

7. Evaluation of the errors of the gnostic estimate of three mean data values.

The Ideal Gnostic Cycle of the given data is identified and optimized in this way.

Distinguishing between the additive and multiplicative data character (step 1 of the analysis) is of a fundamental importance although it is not very popular among statisticians who are using only the Euclidean geometry. Each item of the multiplicative Abelian group can be thought of as the exponential value of an element of the additive group. This multiplicative operation enables to use the operations (2.9), (2.10), (2.15) and (2.16) making possible the application of hyperbolic and trigonometric functions for the basic parameters of the uncertainty.

In the case of a negligible difference between the gnostic and statistical results the conclusion is: very small uncertainty, the robust method is not necessary.

The gnostic evaluation of amount of uncertainty is done by the difference $f_J - f_I$, the residual entropy.

Recall that the statistical mean value is also satisfying a condition of optimality, because it is obtained as the minimum of the sum of squared errors.

16.2 Scale Parameter

They are three kinds of scale parameters in gnostics:

1. the unique **global** scale parameter estimated by minimization of the Kolmogorov-Smirnov test of agreement of the estimated distribution function with the empirical distribution function,

2. the **local** scale parameter estimated by applications of three geometries Q, E or R,

3. **free** scale parameter.

The global and local scale parameters are objectively estimated from data while the free one is set subjectively by the analyst. It is not a data violation but the opposite: it is an instrument in data exploration used to reveal some data hidden data features like in marginal analysis by using the local distribution functions or in analysis of general data relations.

The global scale parameter is optimized along with the bounds of data support in estimation of the global distribution function. The local scale parameter for the geometry Q is obtained by solution of the equation

$$(\pi \cdot S(Z)/2)/\sin(\pi \cdot S(Z)/2) = f_J \qquad (16.4)$$

where Z is the estimated true multiplicative data value and f_J the mean quantifying entropy of the data set. This equation expresses the equality of the scales of continuous with discrete form of a distribution. The value for the geometry E is estimated from an analogous equation and that for the geometry R from the both cases.

16.3 More on the Gnostic Data Aggregation

There is a generally accepted symbol in use for the operation of summarization of data items x_i of a data sample X written as $\Sigma(X)$. The statistical data mean of X is thus $\Sigma(X)/n$ where n is the length of the sample X. It is an inspiration for introducing a special function for the operations of gnostic data aggregation $Gmean(X)$. It is not a function of the type $R^n \to R^1$ like the operation Σ because its value is a data structure consisting of two lists[1]. The first list is a vector of eleven estimates:

1. Gmean(X)[[1]][1] = Mst
2. Gmean(X)[[1]][2] = madX
3. Gmean(X)[[1]][3] = GMQ
4. Gmean(X)[[1]][4] = madQ
5. Gmean(X)[[1]][5] = GMR
6. Gmean(X)[[1]][6] = madR
7. Gmean(X)[[1]][7] = GME
8. Gmean(X)[[1]][8] = madE
9. Gmean(X)[[1]][9] = f_J
10. Gmean(X)[[1]][10] = f_I
11. Gmean(X)[[1]][11] = REE

The first item is thus the statistical mean value of the data sample X. Items $Gmean(X)[[1]][3]$, $Gmean(X)[[1]][5]$ and $Gmean(X)[[1]][7]$ are gnostic means obtained by optimization procedures 16.1, 16.2 and 16.3. Items $Gmean[[1]][2]$, $Gmean(X)[[1]][4]$, $Gmean(X)[[1]][6]$ and $Gmean(X)[[1]][8]$ are mean absolute deviations of median of X, Fq, Fr and Fe. Items $Gmean[[1]][9]$ and $Gmean[[1]][10]$ are both entropies and $Gmean[[1]][11]$ is the relative residual entropy REE. This the relative evaluation of the amount of uncertainty. It is obtained as the ratio

$$REE = (f_I - f_J)/(f_I + f_J) \tag{16.5}$$

---
[1] The terminology of the R-language (www/r-project) is applied here along with its symbolics: The square brackets denote items of a vector or of a matrix, the double square brackets identify items of the lists.

where f_I and f_J are mean values of both entropies. The quantity REE is thus a number from the interval $(0,1)$.

The second list of *Gmean*'s values is a $L \times 6$ matrix with the following columns (i = 1,..,L):

1. Gmean(X)[[2]][i,1] = X[i]
2. Gmean(X)[[2]][i,2] = Fq[i]
3. Gmean(X)[[2]][i,3] = Fr[i]
4. Gmean(X)[[2]][i,4] = Fe[i]
5. Gmean(X)[[2]][i,5] = p[i]
6. Gmean(X)[[2]][i,6] = IQe[i]

where X is the sample, Fq and Fe are the estimates of bounds of uncertainty domains obtained by the second axiom and Fr their additive or multiplicative mean. The $p[i]$ is the probability of the estimate $Gmean(X)[[1]][7]$ and the $IQe[i]$ is the information (8.12) divided by log(2) born by the data item $X[i]$. Item $Gmean[[1]][2]$ is the standard deviation of the statistical mean's estimate. Items $Gmean(X)[[1]][4]$, $Gmean(X)[[1]][6]$ and $Gmean(X)[[1]][8]$ are obtained as mean absolute deviations of the medians of functions Fq, Fr and Fe.

There is a special way of computing the variables Fq and Fe by using the second gnostic axiom: Assume that the values of variable X are members of the Abelian multiplicative group, otherwise it would be necessary to transform its values by the exponential function. Quantifying and estimating residuals $f_J(X_i)$ and $f_I(X_i)$ are hyperbolic and trigonometric cosines. Aggregating them by the second axiom creates quantities $\sum_i^L (f_J(X_i))$ and $\sum_i^L (f_I(X_i))$. Attaching the aggregated values to individual data values is thus equivalent to multiplying each of i-th data value by weight $wq_i = X_i / \sum_i^L (f_J(X_i))$ and/or $we_i = \sum_i^L (f_I(X_i))$. Variables Fq and Fe are then formed as a series of values $X_i * wq_i$ and $X_i * we_i$. Variable $Fr[i]$ is then an arithmetical or multiplicative mean of $Fq[i]$ and $Fe[i]$ (according to the geometry 'Q' or 'E').

The variable $p[i]$ is the probability density of the i-th data item and $IQe[i]$ is its improbability.

16.3.1 Example

TABLE 16.1
List 1 of the Gmean(stackloss[,4])

Mst	madX	GMQ	madQ	GMR	madR	GME	madE	fJ	fI	RRE
17.52	5.93	15.41	5.75	15.15	5.76	14.68	6.76	1.14	0.90	0.12

The second list of results is also interesting:

TABLE 16.2
List 2 of the Gmean(stackloss[,4])

Sample	Fq	Fr	Fe
42	20.948	14.625	10.211
37	19.085	14.645	11.237
37	19.085	14.645	11.237
28	16.035	14.703	13.481
18	13.710	14.836	16.054
18	13.710	14.836	16.054
19	13.844	14.817	15.858
20	14.009	14.800	15.635
15	13.550	14.900	16.384
14	13.608	14.925	16.369
14	13.608	14.925	16.369
13	13.742	14.951	16.267
11	14.322	15.008	15.727
12	13.971	14.980	16.060
8	16.560	15.100	13.770
7	17.984	15.132	12.731
8	16.560	15.101	13.770
8	16.560	15.101	13.770
9	15.551	15.069	14.603
15	13.551	14.900	16.384
15	13.551	14.900	16.384

The means Mst, GMQ, GMR and GME differ. The estimates Fq, Fr and Fe exhibit other worthwhile features:

1. mad(Fq) = 5.75
2. mad(Fr) = 5.76
3. mad(Fe) = 6.76

What is shown by these relations is important: the estimates Fq, Fr and Fe could be used as good filtered values of the sample stackloss[,4]. They will differ one of each other by their robustness which results from the geometry applied in estimation.

To apply the gnostic aggregation, the function $Gmean$ is run twice, at first for the homoscedastic case (with the parameter $varS = F$ while the second run is done with $varS = T$ for the heteroscedastic case. Each run is generating four cases of the double $mean(x), error(x)$ for the geometries S, Q, E and R where the S states for the statistical means. To determine the most suitable geometry, the quality of results is ordered to minimize the ratio $|error|/mean$.

16.3.2 Example of Robust Estimation of the Mean of Multiplicative Data

Some examples of application of the function Gmean to real data could support the theoretical statements. To start return to already cited data called NIST37 [39] which were shown to be very precise and then to the much worse data set NIST12 [38]. The analysis gives following results where the ± is the error of the mean's estimate:

TABLE 16.3
Estimates of the means of the NIST37 data

Method	Mean ± Error
GMR	334.3132 ± 0.0741
Stat.	334.3132 ± 0.0741

The values of estimated means by both approaches are the same as could be expected because of high precision of the measurements. The filtered data are all equaling each other. The errors in Tab. 16.3 correspond to the high precision of data. Mean absolute deviation has been applied for comparability of the errors. The gnostic estimate of the true value is close to bounds of the resulting kernel matrix obtained as Tab. 15.6. Table 16.3 documents that the gnostic estimate tend to the classical ones when the uncertainty is very weak.

Another picture is obtained with matrix NIST12 (Tab. 16.4).

TABLE 16.4
Estimates of data NIST12

Method	Mean ± Error
Gnos.	410.2792 ± 1.4085
Stat.	410.2792 ± 1.4085

The statistical case is here equal to the best gnostic result. Both estimating errors are here much larger. The cause is already known, the data are non-homogeneous. The gnostic estimate does not correspond here to the results of data calibrating because of the non-homogeneity of sample NIST12 which makes the estimate biased. The results partially improve after leaving out the two main outliers Nos. 5 and 9 to make the sample homogeneous:

TABLE 16.5
Estimates of data NIST12H

Method	Mean ± Error
Gnos.	429.3514 ± 1.1069
Stat.	429.3200 ± 1.1120

However, failing of the statistical mean value can be shown by other real data loaded with a higher variability. The daily concentration of ozon measured in New York, May to September 1973, particularly the mean ozone in parts per billion from 13.00 to 15.00 hours at Roosevelt Island (Tab. 16.6).

TABLE 16.6
Estimates of ozon concentration in New York

Method	Mean ± Error
Gnos.	39.0059 ± 13.5695
Stat.	42.0991 ± 25.2042

16.3.3 Robust Estimation of the Mean of Simulated Data

To demonstrate the application of the same approach to additive data having different distribution functions the simulated data are to be applied. Two cases will be considered:

Normally distributed data are in R-system generated by command $rnormal(n, mean, sd)$ where n is the amount, $mean$ the mean value and sd the standard deviation of data to generate.

Cauchyan data are analogously generated by command $rcauchy(n, location, scale)$.

The case of normal data rnorm(50, 2, 10):

TABLE 16.7
Estimates of normal data $rnorm(50, 2, 10)$

Method	Mean
Gnos.	2.1175
Stat.	2.3910

The case of Cauchyan data:

TABLE 16.8
Estimates of Cauchyan data $rcauchy(50, 2, 10)$

Method	Mean ± Error
GME	679.6471
Stat.	352.7160

This data set was additive and the most suitable geometry was 'E,' therefore the estimate of the mean *GME* has been applied.

To show the application to the real additive data, the financial ratio Total Return (TOTR) will be used which is the year change of the share price plus year sum of dividends paid divided by the total assets. The values of TOTR of the IBM in years from the 4-th quarter 1996 through the first quarter 2000 are in Tab. 16.10.

TABLE 16.9
Total Return values of the IBM Company

Year	TOTR
96_4	0.21967871
97_1	−0.09174917
97_2	0.31803279
97_3	0.17673130
97_4	−0.01108491
98_1	−0.00525687
98_2	0.10740794
98_3	0.12113717
98_4	0.43653696
99_1	−0.03745085
99_2	0.45974612
99_3	−0.06290135
99_4	−0.10747934
00_1	0.09497103

Estimation of means of the Total Return:

TABLE 16.10
Estimates of TOTR of the IBM Company

Method	Mean
GME	0.1210
Stat.	0.1156

The errors of gnostic estimates appeared to be in some cases better than the statistical means, especially at large uncertainty. This is because the gnostic means belong to the optimized Ideal Gnostic Cycle and because of the robustness of the estimates.

16.4 Conclusions

Estimation of a mean value is a most frequently asked task of data treatment: to represent an one-dimensional data sample by a single number. There are two key problems in solving this task, the weight of each individual data item and the way of aggregating the data weights of all data.

The uniform weight applied by calculating the statistical mean value suffers from the non-robustness caused by strong differences between data values. Application of the weights proportional to the reciprocal variances of the data sub-samples requires availability of cluster analysis of data sample and does not ensure maximal information of the results, because the weights given to individual data items do not correspond to the item's own uncertainty but to the "collective" uncertainty of the sub-sample. As results from the mathematical gnostics, the maximum of result's information is reached by data weighing by individual weights dependent on the individual uncertainty of the data item.

It also results from the gnostic theory that the additive aggregation of the data and their powers and products applied in statistics is applicable only in cases of a weak data uncertainty. The right and more universal way of data aggregation is additive aggregation of gnostic entropies of data which are non-linear data functions. The results obtained in this way are more precise than in the statistical case even when measuring their precision by the statistical way.

17
Measuring the Data Uncertainty

17.1 Shortly on the Standard Approach

NIST (National Institute of Standards and Technology) is a world-known and respected American institution founded in 1901. Its task is to promote U.S. innovation and industrial competitiveness by advancing measurement science, standards, and technology in ways that enhance economic security and improve the quality of life. The Standard Reference Data of NIST must meet stringent evaluation criteria[1] by the following methods:

1. Uncertainty assignment

   Original source

   NIST, sometimes based on internal consistency of all available data

2. Use of standards

   Instrument calibration

   Other standards, e. g., internal standards, wavelength standards

3. Comparisons

   Agreement between experimental data and data calculated by various methods, e. g., ab initio models, semi-empirical methods

   Agreement with data obtained by independent measurement methods

   Agreement with other reliable data

4. Consistency within or between data sets

   - Follow known laws or rules of nature, e. g., positive pressures
   - Obvious errors
   - Self-consistency adjustment, e. g., entire data set
   - Empirically corrected, e. g., known to be off by a certain percentage

---
[1] www.nist.gov/srd/critical-evaluation-criteria

- Re-analysis to maximize consistency of data from various sources

5. Trends
   - Regularity of empirical trends
   - Fit of data to systematic trends
   - Large number of outliers

6. Human element
   - Expert judgment
   - Stature of providers of external data
   - Misprints in publication

7. Data generation
   - Inclusion of metadata
   - Documented procedures
   - Independent variables identified and controlled

The detailed methods of evaluating and expressing the uncertainty are described in [100]. These guidelines are based on methods formulated by the CIPM (International Committee for Weights and Measure). There are two distinguished components of uncertainty: **A**—those which are evaluated by statistical methods and **B**—those which are evaluated by other methods. In agreement with CIPM, the uncertainty in category **A** is represented by a statistically estimated standard deviation s_i equal to the positive square root of the statistically estimated variance s_i^2. A type **B** evaluation of standard uncertainty is usually based on scientific judgement using all the relevant information. The quoted uncertainty which is a stated multiple of standard deviation is recommended to convert to the standard uncertainty by dividing it by the multiplier. The normal distribution is then applied to determine the bounds $a_{(-)}$ and $a_{(+)}$ such that the probability of the quantity to fall into the interval of bounds is 1/2. Then the standard uncertainty is $u_j = (a_{(+)} - a_{(-)})/2$. Repeat this operation for the probability 2/3 to get another standard uncertainty u_j. The third value of standard uncertainty $u_j = ((a_{(+)} - a_{(-)})/2/sqrt3$ is obtained by modeling the probability as uniformly distributed over the interval from $a_{(-)}$ through $a_{(+)}$ covered "practically" 100% of cases. If the distribution to model the quantity is rather triangular than rectangular, then $u_j = a/sqrt6$. In the case of using the normal distribution, the "almost all" of the possible values (99.73%) fall into the interval from $a_{(-)}$ to $a_{(+)}$ equaling to three standard deviations. The standard uncertainty is then $u_j = (a_{(+)} - a_{(-)})/2/3$. The *combined standard uncertainty* u_c can be obtained by the *law of propagation the uncertainty* or RSS method (square root of sum of squares) from several sources of uncertainty. Another case of combined uncertainty is the application of correction factors to compensate for each recognized systematic effect. There also is the *expanded uncertainty* U obtained by multiplying the uncertainty u_c by a *coverage factor* k which is typically in the range 2 to 3. The coverage factor is chosen on basis of confidence to be associated with the uncertainty U.

17.2 The Need of Objective Measuring the Variability

It is obvious that the standard approach to determination of measuring variability is rather subjective and that it does not correspond to the declared requirements. The difference between two components of data variability (the variability of the true data items and uncertainty of its measuring) is not taken into account. The role in establishing the uncertainty level is more on the analyst than on the observed data. The mathematical gnostics comes with an objective evaluation of data variability based on observed data and on their thermodynamic nature.

Three entropies in gnostic theory are connected with the Ideal Gnostic Cycle of a single data item: E_J (8.2) evaluating the entropy's increase in quantification, E_I characterizing the entropy's decrease in estimation and their difference $f_J - f_I$, the residual entropy. The entropy E_J was shown to equal $f_J - 1$ where f_J is additive because of being isomorphic image of the energy of a relativistic particle, which is additive. The entropy E_J is therefore also additive. The entropies f_I and $f_J - f_I$ are additive due to the second gnostic axiom which is supported by the Energy and Momentum Conservation of relativistic mechanics. It means that the mean entropy of a data sample can be obtained as the mean of entropies of the individual data. We already know how to estimate the variables f_J and f_I. This enables measuring of the data uncertainty by estimating the means of entropy and not by estimating the means of "uncertainties" of data like it is prescribed by the *CIPM*.

17.3 The Triplication of the Mean Values

The statistical mean value of a data sample obtained as the additive mean of data items minimizes the variance of the sample—which are taken as the uncertain variability. The bounds of data values are then obtained like the mean plus/minus the probable value of variability. When using the gnostic approach, we estimate the bounds of variable values as functions of entropies as discussed in previous chapter. The upper bound is determined by the minimization f_J, the lower by the maximization f_I and the "best" value by the minimization of the residual entropy $f_J - f_I$. The upper bound of the mean of an variable sample can be called GMQ because it is determined by the quantification phase of the IGC. The lower bound will be analogously GME and the "best" mean value GMR. The dependence of these mean values on entropies leads thus to three tasks:

1. to find GMQ minimizing the f_J,
2. to find GME maximizing the f_I,
3. to find GMR minimizing the $(f_J - f_I)$.

Instead of one (statistical) mean value we have three, but with different interpretation not only of what is called *the mean* but also of *the error of the mean*. We are obtaining two sharp bounds GMQ and GME delimiting the extremes of variability of the estimate GMR of the mean value. The absolute uncertainty is obviously estimated as $GMQ - GME$ (the residual entropy of the Ideal Gnostic Cycle).

However, the choice of possible geometries is broader than three. We already know, that the gnostic characteristics converge to the statistical ones when the uncertainty sufficiently decreases. This makes necessary to extend the number of potential geometries to four by inclusion of the statistical geometry. But the data aggregation is dependent on the data weights determined by scale parameters which may be constant for the whole data sample (the homoscedastic data) or individual in dependence on the data item's uncertainty (the heteroscedastic data). The number of potential geometries thus extends to four. The choice of the most suitable geometry for analysis of a real data sample is left to data as shown in chapter Robust Estimation of a Constant.

17.4 The Need of a Unit of Uncertainty

Why do we need the uncertainty measurements? The primary question is why do we need measuring: to get values of quantities. However, each measurement involves more or less uncertainty. No measurement is complete without estimating its plausibility or its probable error. The cost of measuring devices is dependent on the devises' precision and stability. The final judgment on each product's quality can be obtained by its operation and by measuring and evaluation of results' variability and uncertainty. But data samples may originate not only in measuring. Many kinds of activity of men result in data requiring to be analyzed and their quality to be evaluated. It is necessary to compare the results of this activity not only by their values but also by undesirable variability of values evaluated by a reliable method. It is a standard practice to present the quantitative results of a process in the form of the mean value plus/minus the estimated error. Both these quantities—when

The Need of a Unit of Uncertainty

obtained as in statistics—can suffer of non-robustness and uncertainty and may involve errors by improper aggregating of their elements. Measuring of their variability by gnostic entropy is the chance for significant improvement. To measure the uncertain data component of variability would naturally require separation of both component but quantification of uncertainty is the necessary first step. The problem of such separation will be considered in chapter Filtering.

When deciding to measure the data uncertainty by means of the residual entropy, it is natural to think of the measuring unit to be applied. The measuring unit of thermodynamic entropy is the ratio Joule/Kelvin. It is an extensive quantity proportional to the size of the system. Because it is impractical to give the system size along with the extensive quantity, the physical chemistry uses the molar quantity and in technical applications the measuring quantity related to kilograms. Measuring units related to the volume are in use in some applications. To get a dimensionless unit one can use the ratio of heat in calories and temperature expressed in Rankin's degrees divided by the universal gas constant. However, in the notion of gnostic entropy, both heat and temperature are virtual quantities obtained as images of a number by a gedanken-experiment, there is neither real heat nor temperature. To measure the data uncertainty, the thermodynamic measures are evidently of no use. But there is one additional requirement to measuring unit of uncertainty: to be sensed by a human, to give him/her a chance to feel the actual size of uncertainty.

Such a chance appeared already in the forgoing chapters where the gnostic estimates of mean values of quantities were presented as functions of f_J and f_I. Write the mean values of the observed quantity X as $E(f_J)$ and $E(f_I)$. The f_J is the hyperbolic cosine by (2.9) and the f_I is a trigonometric cosine by (2.15). Therefore the following equations hold:

$$E(f_J) = \mathrm{mean}(\cosh(X)) \qquad (17.1)$$

and

$$E(f_I) = \mathrm{mean}(\cos(X)) \ . \qquad (17.2)$$

Instead of estimating the functions $\cosh(X)$ and $\cos(X)$ and then using these equations we found the means $E(f_J)$ and $E(f_I)$ as GMQ, GME and GMR in the previous chapter by optimizing the Ideal Gnostic Cycle. As the best mean we shall use the quantity providing with the best results.

We know that the difference $f_J - f_I$ is the residual entropy. The difference $GMQ - GME$ of mean values is thus an image of the residual entropy measured in the same units as the means, that is the same unit as each of the individual data items. When reporting the measurement result of the variability using the units of the data one can use the dimensionless form of $GMQ - GME)/(GMQ + GME)$ called the *relative residual entropy*: the image of residual entropy divided by the total of entropy changes within the Ideal Gnostic Cycle.

The sensible dimensionless value of relative amount of uncertainty can be thus defined as

$$RRE = (E(f_J) - E(f_I))/(E(f_J) + E(f_I)) \qquad (17.3)$$

This expression is worth to be commented. It gives the image of the relative value of estimated means of residual entropy, it is a relative measure of the uncertainty. When there is no uncertainty, both $E(f_J)$ and $E(f_I)$ are equal to 1 and the estimating efficiency is zero, there is nothing to estimate. However, with rising uncertainty the ratio $E(f_I)/E(f_J)$ tends to 0. The maximum relative uncertainty corresponds to the value $REE = 1$.

The operation of data cleaning considered in the chapter Filtering removes only the uncertainty impacts and enables thus the changes of true data values to be separated and explained. Till this chapter, the assumption of zero variability of the true values of data will be accepted.

17.5 The Error of a Mean

To apply different methods of estimating the mean value's error, we need a method evaluating the error component of the estimated values. We can use the thermodynamic evaluation by means of variables f_J, f_I or the residual entropy $f_J - f_I$ and its relative value RRE, but the result would be hardly comparable with the statistical measure. We are therefore using an inspiration of statistical mean absolute deviation from the median by using its generalization of the form of

$$mad = mad(Fy) \ . \qquad (17.4)$$

where mad(.) symbolizes the operation mean absolute deviation from the median. Symbol y states for variables x (X), q (Q), r (R) or e (E) and Fy are Fq, Fr or Fe, variables Fy are estimations connected with the solutions of equations (16.1), (16.2) and (16.3) discussed in chapter "Robust Estimation of a Constant."

17.6 Examples

All examples in this chapter are prepared by the function Gmean applied twice to include both homo- and heteroscedastic data and choosing the most suitable from eight geometries.

Examples 187

17.6.1 Swiss Fertility and Socioeconomic Indicators (1888) Data

This data matrix is taken from the database of the R-project (www.r-project.org). The data form a matrix with 47 observations on 6 variables, each of which is in percentage, i.e., in [0, 100].

TABLE 17.1
Swiss Fertility and Sociometric Indicators

Variable	Description
Fertility	Ig, common standardized fertility measure
Agriculture	% of males involved in agriculture as occupation
Examination	% of draftees receiving highest mark on army examination
Education	% of education beyond primary school for draftees
Catholic	% 'catholic' (as opposed to 'protestant')
Infant Mortality	% of live births who lived less than 1 year

All variables but Fertility give proportions of the population. Results are in Tab. 17.2 in the order of *RRE*.

TABLE 17.2
The means and uncertainty in Swiss Data

Variable	Col.	Mst	madst	Geom.	varS	GM	madG	RRE
Catholic	5	41.14	18.65	R	TRUE	80.19	11.12	0.407
Examination	3	16.49	7.41	St	FALSE	16.49	7.41	0.246
Fertility	1	70.14	10.23	R	FALSE	71.07	10.02	0.206
Agriculture	2	50.66	23.87	E	FALSE	56.61	24.54	0.199
Education	4	10.97	5.93	Q	TRUE	8.16	3.65	0.183
Inf. Mortal.	6	19.94	2.82	Q	TRUE	19.84	2.67	0.089

Variable Mst is the statistical mean value, *madst* its mean absolute deviation, Geom. the most suitable geometry, varS the setting of the scale parameter, GM the gnostic estimate of the mean value obtained by proper aggregation when evaluated in geometry Geom., madG its mean absolute deviation and RRE the relative residual entropy. The lines of the table are in order of relative residual entropy *RRE*. As seen in Tab. 17.2, the variabilities in different columns of the same matrix may be very different. The differences between the statistical and gnostic values do not depend only on the value of variability, but also on the data distribution. The error caused by different aggregation in the mean value is nearly negligible in the case of Examination and Infant. Mortality, but in the case of variable Catholic it reaches a considerable part of the measured value. But let us have a look at its value $Ms \pm madX$ which is ±41.14. What it really says? That the mean value can be expected anywhere between $41.14 - K \cdot 18.65$ and $41.14 + K \cdot 18.65$ where the multiplier K may be 2 or 3. If we would know the distribution function of the variable, we could estimate the effect of variability on the mean in terms of probability. But we generally don't have a reliable distribution in statistics. When considering the gnostic result in the similar form of 80.19 ± 11.12 we know exactly what it means: the variability can move the mean "in the mean of 47 observations" from GME to GMQ which represents this time the interval MGQ, MGE, that is $(18.64, 80.19)$. This is only a small sub-interval of the statistical interval. (If we would like to know what it is in terms of probability, we could compute it too.)

17.6.2 Financial Statement Analysis

Data for this example are taken from the database Compustat. Financial ratios were calculated from the items of financial statements of IBM for the fourteen quarters of year from the 1966/4 till 2000/1. The ratios are defined in Tab. 17.3 where following symbols are used: TA—total assets, TL—total liability, CA—current assets, CL—current liabilities, $EBIT$—earnings before taxes, NS—net sales, dD—dividends for the lagged year, P—stock price, dP—year change of stock price, E—earnings. The ratios are given as functions of these items. Tables 17.3 and 17.4 are ordered by the variability.

TABLE 17.3
Basic financial ratios of IBM

Ratio	Name	Formulae
TOTR	Total return	(dP+dD)/TA
RWC	Relative working capital	(CA-CL)/TA
ROA	Return on assets	EBIT/TA
TATO	Total assets turnover	NS/TA
P/E	Price/earnings ratio	P/EBIT
PS	Price of stock	P
RTL	Financial leverage	TL/TA
RCL	Relative current liability	CL/TA

The lines in Tab. 17.4 are ordered by the uncertainties RRE which are somewhat different. This may be interpreted by the bad estimate of its state (PS and P/E) and a non-quiet behavior of the market ($TOTR$). The variability of the RWC may be explained by its components (current assets and current liabilities) which change as a daily routine within the financial control of the business. The differences between statistical and gnostic estimates of the variable TOTR and of their errors document the impact of the way of data aggregation. It may result in problems of prediction which is inevitable for planning and budgeting.

Examples

TABLE 17.4
Uncertainty in basic financial ratios of IBM

Ratio	Col.	Mst	madst	Geom.	varS	GM	madG	RRE
PS	6	75.225	36.671	Q	FALSE	68.620	23.700	0.264
TOTR	8	0.116	0.191	Q	FALSE	0.121	0.182	0.246
P/E	7	0.014	0.004	Q	FALSE	0.013	0.004	0.241
CLdA	1	0.415	0.026	Q	FALSE	0.414	0.026	0.221
TATO	2	0.254	0.022	Q	FALSE	0.254	0.021	0.211
EBITdA	4	0.030	0.005	Q	FALSE	0.030	0.004	0.207
RWC	3	0.077	0.014	R	FALSE	0.083	0.015	0.194
TLdA	5	0.759	0.010	R	TRUE	0.761	0.010	0.090

There is also a warning in these results for three important economic professions, accountancy, financial management and auditing. The accounting applies additive aggregation of items, managers use the aggregated figures to control the business and auditors judge and confirm the results of statements which summarize the partially false numbers as shown in Tab. 17.4. The warning sounds: when summarizing the data, take into account their uncertainty! Additive aggregation of accounting items is the natural way of accumulating money and is necessary for the primary role of the accountancy, to trace movements of each penny of the property. However, the real "power" of a sum of money is dependent on its size and on its stability. The evaluation of uncertainty would be thus a welcome complement to the sums or means of the money. It could be worth of enriching the financial statements documentation by a report of statements' variability and uncertainty which would help to evaluate the stability of the particular business.

17.6.3 Weather Parameters

Data for this example are taken from the matrix *airq* that is in the database of the R-project. They characterize the weather in New York City from May 1 through September 30 (1937).

TABLE 17.5
Variables of weather parameters

Variable	Units
Solar Rad.	lang
Ozone	ppb
Temperature	°F
Wind	mph

The results of analysis are in Tab. 17.6 ordered by the uncertainty *RRE*.

TABLE 17.6
Uncertainty in weather parameters

Variable	Col.	Mst	madst	Geom.	varS	GM	madG	RRE
Ozon	1	42.13	32.99	Q	FALSE	28.42	1.33	0.309
Solar Rad.	2	185.93	90.06	R	TRUE	144.35	18.44	0.0.257
Wind	3	9.94	3.41	Q	TRUE	9.37	2.88	0.128
Temp	4	77.79	10.38	R	TRUE	78.48	10.41	0.100
Ozon	1	42.10	25.20	Q	FALSE	30.01	13.57	0.296
Solar Rad.	2	184.80	91.92	R	TRUE	219.73	103.56	0.161
Wind	3	9.94	3.41	Q	TRUE	9.37	2.88	0.128
Temp	4	77.79	10.38	R	TRUE	78.48	10.41	0.100

The variability of these parameters is of course caused rather by natural behavior of the variables than by measuring errors. The largest uncertainty is that of the *Ozone*. However, even the smaller uncertainty of other parameters causes the not negligible differences between the statistical and gnostic mean values, for instance in the case of *Solar Rad.*

17.6.4 An Important Medical Parameter

Dataset *lh* (the luteinizing hormone in blood sample) can be also found in the database of the r-project. This is a hormone produced by gonadotropic cells in the anterior pituitary gland of females. It controls among others the process of ovulation. Before evaluating the variability of the sample *lh* of 48 values, we check the homogeneity by means of the distribution function. As seen in Fig. 17.1, the sample is homogeneous, there is only one local maximum on the density function over the infinite data support.

The results of uncertainty analysis of the data sample are shown in Tab. 17.7.

TABLE 17.7
Uncertainty in the lh hormon

Varbl	Col.	Mst	madst	Geom.	varS	GM	madG	RRE
lh	1	2.400	0.593	R	TRUE	2.338	0.593	0.116

Examples 191

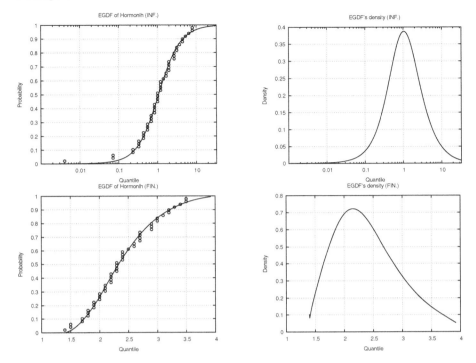

FIGURE 17.1
The homogenous probability distribution function of hormon lh

The measurements are in this case relatively precise, therefore the small difference between *Mstat.* and *MGR*.

17.6.5 Non-homogeneous Data

It is worth of returning to the non-homogeneous data from the database NIST analyzed already in connection with the thermodynamic analysis. Looking at the global distribution function of homogenized sample NIST12H (window (1,2) of the Fig. 15.1), one sees that—although the sample is tested as homogeneous, its structure insinuates existence of inner sub-samples. To see the finer structure of the sample, we apply to the same data the local distribution function receiving Fig. 17.2. We are looking for a homogeneous sub-sample. Therefore we play with the scale parameter of the local distribution till we get for the value $S = 1.5$ four peaks of the density distributions. The second peak seems to be appears to be homogeneous.

Using the numerical output of the density distribution over the infinite data support, we find, that the minima of the second peak delimit the six data values 426.65, 427.05, 427.10, 427.15, 428.00 and 428.15. To verify the

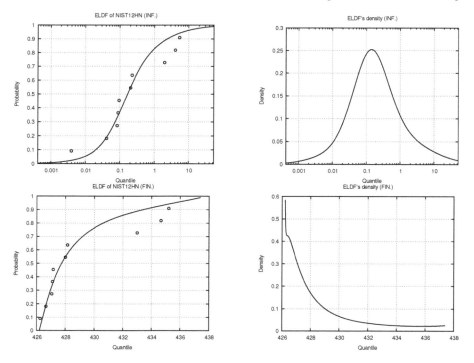

FIGURE 17.2
The local distribution function of the data of NIST12H

suspected non-homogeneity of this sub-sample, we compute the global distribution function of it. Result is in Fig. 17.3.

The peak appears to be non-homogeneous but three inner data 427.05, 427.10 and 427.15 might form a homogeneous sub-sample. To check it, we apply to them again the global distribution function and obtain Fig. 17.4.

The peripheral data 427.05 and 427.15 appeared to be not far enough from the central value 427.10 to form with it a homogeneous data sub-sample. The conclusion is, that sample NIST12 has just one homogeneous sub-sample NIST12H.

It can be interesting how the uncertainty of the data sub-samples was decreasing during elimination of outliers in Tab. 17.8.

Two methods of eliminating the outliers were applied:

1. Leaving out the extreme value,
2. Reweighting by successive substitution of a posteriori data weight instead of the a priori ones.

The following experience results from this analysis:

Examples

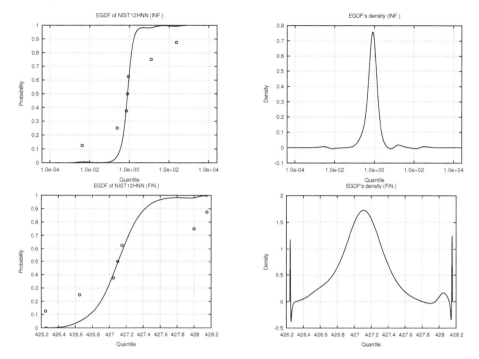

FIGURE 17.3
Global distribution function of the data of the large peak in NIST12HN

1. There can be a large error in statistical mean value caused by improper aggregation of uncertain data dependent on their variability.
2. The non-homogeneity of the probability distribution must not be overlooked in data analysis. It may falsify the mean value and increase the result's uncertainty considerably.
3. Even the inner structure of a homogeneous sub-sample is worth of attention because its analysis can be useful for understanding of the mean value.
4. The homogenization by elimination of outliers may enable a better information quality than the reweighting to be obtained although of using less—but more suitable—data.
5. The elimination of outliers leads automatically to a decrease of uncertainty as shown by comparison of values RRE of cases $NIST12$ and $NIST12H$.

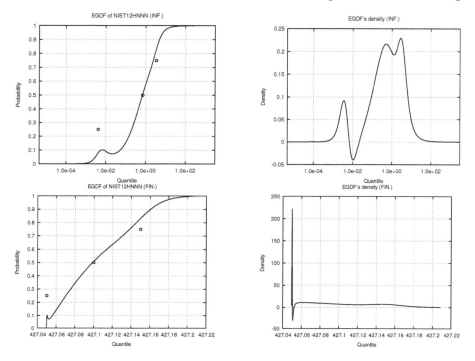

FIGURE 17.4
Global distribution function of the data of the three inner data in NIST12HNN

TABLE 17.8
Uncertainty of data NIST12 during analysis

Variable	Col.	Mst	madst	Geom.	varS	GM	madG	RRE
NIST12	1	410.28	1.41	Er	TRUE	425.84	1.41	0.0491
NIST12H	1	429.32	1.11	Q	FALSE	429.35	1.11	0.0463
NIST12HN	1	427.35	0.40	Q	FALSE	427.35	0.41	0.0013
NIST12HNN	1	427.10	0.07	St	FALSE	427.10	0.07	7.0E-06

17.6.6 Parameters of Uncertainty

The last point should not cause a misunderstanding: the presence of outliers in a sample is a disorder and their leaving out should improve the sample—at least as one would intuitively expect. Let us consider this case in more detail by using other parameters of the sample's uncertainty in Tab. 17.9:

The symbol $E(x)$ denotes here again the additive mean of the variable x. The other symbols are:

1. f_J ... quantifying uncertainty of a data item, (2.9)
2. f_I ... estimating uncertainty of a data item, (2.15)

Examples

TABLE 17.9
Thermodynamic parameters of *NIST12* and *NIST12H*

Data	$E(f_J)$	$E(f_I)$	RRE
NIST12	1.0931	0.5934	0.2963
NIST12H	1.0427	0.7542	0.1605

3. *RRE* ... relative residual entropy.

The mean values of the quantities can be determined because the theory provides formulae for individual data items. It is shown by the Tab. 17.9, that the residual entropy $E(f_J) - E(f_I)$ of sample *NIST12H* does not exceed the value of *NIST12*. The behavior of all individual components of the gnostic characteristics in the case of leaving out outliers is strange enough to make their details interesting in Tab. 17.10:

TABLE 17.10
Individual data information of *NIST12* and *NIST12H*

NIST12	I(NIST12)	I(NIST12H)
435.20	0.08869	0.10136
433.00	0.08827	0.10086
427.10	0.08713	0.09948
428.15	0.08733	0.09973
322.00	0.06393	Left out
427.15	0.08715	0.09950
434.65	0.08859	0.10124
426.25	0.08696	0.09929
308.15	0.06050	Left out
427.05	0.08712	0.09947
428.00	0.08730	0.09970
426.65	0.08704	0.09938

The numbers *I(.)* represent the contribution of the data item to the full information of the column which is 1. The values of *I(NIST12)*'s data 322.00 and 308.15 represent the least information values 0.06393 and 0.06050, they are obvious outliers. However, their leaving out increases the information born by all other items of *NIST12H* so that the mean information of the sample becomes the same (full, 1) as the mean information of *NIST12*.

There are two general lessons from this example:

1. The check of the non-homogeneity of a sample is a must of the data analysis. It relates especially—although not exclusively—to medicine where a generally accepted way of determination of the s.c. **reference values** of health parameters exists by the mean

values of 95% of measurement done on "healthy" people without checking the homogeneity of the parameters of participants.

2. Evaluation of uncertainty of data may happen to become a second must of the data analysis.

17.7 Discussion on Different Means

The gnostic statement on necessity of non-linear aggregation of uncertain data may seem to be strange and incomprehensible because of being in a conflict with experience of a human. Even as a baby is everybody used to cover distances by repeatedly added steps and to spare money by adding the spared penny to a cash. However, there are differences between steps due to their difficulty caused e. g. by decline of the way. This is the reason why distances between places are given in the High Tatras mountains in Slovakia not in kilometers but in time a mean walker needs to pass. Or another example: compare the costs of covering the distance of a step by different transport means. The cheapest way is probably the walking, but otherwise the comfort of the movement and its velocity enter as important factors of the cost. The distance can be thus seen as a function of several variables and its evaluation can be different in dependence on the aspects acting as metrics.

Even the "classical" way of expressing a variable quantity in the form of mean ± STD assumes that a value is at least two-dimensional function at which its variability determines its quality. This is why the more precise instruments cost more than the less precise ones. People accept the price of variability when betting on some uncertain events or when buying tickets of a lottery. People accept price for insurance when it corresponds to the danger. The invested sum of money must correspond to the estimated variability. It has been shown above how the variability can be evaluated and why the aggregation of variable data must respect the uncertainty. This makes the mean of data values dependent on uncertainty and on the metric of variability evaluation. There is a non-random similarity between the role of uncertainty in influencing the quality of the data mean and the role of thermodynamic entropy in influencing the quality of a thermodynamic cycle because the entropy of uncertainty has been derived from its thermodynamic predecessor. The uncertainty is an evaluation of the quality of the estimate of the properly aggregated mean value.

17.7.1 Re-definition of Variance

Three thermodynamic variables $cosh(Fi)$, $cos(fi)$ and $cosh(Fi) - cos(fi)$ were selected as parameters of the Ideal Gnostic Cycle. Variance is defined in statistics as the expectation of squared deviation from its population or data sample mean. This definition is not as general as wee need. There are following critical points:

1. geometry to measure the deviation is missing,
2. the square deviation is a special and too narrow choice of operation,
3. the mean of statistical data is frequently obtained as the sum of data divided by their number.

The advantage of quadratic function is that its first derivative is a constant. There is a special point—the argument of its maximum frequently used as an optimum. However, all three parameters of the Ideal Gnostic Cycle also have a maximum of their first derivative which is generally different from a constant. On the other hand, as was already shown, thanks to considering account of two geometries with a opposite sign of robustness, one obtains a smooth first derivative—a favorable feature. The parameters of the IGC due to their being in connection with optimality of information can be used instead of variance to characterize the data uncertainty.

17.8 Conclusions

Measuring of data uncertainty is a way of evaluation the quality of the estimate.

The proper aggregation of variable data leads to alternative values of mean values of samples of variable data and to an alternative conception of statistical errors. Instead of standard deviations the bounds of uncertainty of mean values can be used providing a substantially lower estimate of the range of errors. The proper aggregation of data leads to lowering of the errors of mean's estimates. The consequences of the improper data aggregation may lead to degradation of all statistical operations based on application of simple mean sums of data values including estimation of data statistics.

The possible non-homogeneity of data must be taken into account in analysis to escape falsification of results by data variability.

18
Homo- or Heteroscedastic Data

In statistics, a sequence or a vector of random variables is homoscedastic iff all random variables in the sequence or vector have the same finite variance. This is also known as homogeneity of variance. The complementary notion is called heteroscedasticity.

In mathematical gnostics the role of variance is taken over by the more general variables quantifying entropy f_I (2.9) and estimating entropy f_J (2.15) which are tending to the variance with sufficiently decreasing uncertainty. These both are dependent on the scale parameter S and on the parameter Φ, resp. ϕ, the value of which is individual for each data item. The scale parameter S can be a constant for all data of the sample or individual for each of the items. In the former case the data sample will be taken as homoscedastic, in the latter case as heteroscedastic.

18.1 Decision Making

Making decision between homoscedastic and heteroscedastic versions of data analysis is one of key problems in applications because of serious consequences of the decision: results of analysis may significantly change in dependence on the decision. The form of the distribution function is dependent on this decision like results of data filtering and prediction and most of other results of data analysis. There are two ways of the decision making:

1. Subjective.
2. Objective: data conditioned automatic decision making.

Many gnostic algorithms have a Bollean parameter called *varS* which decides between homo- and heteroscedastic cases. Its value can be set subjectively when testing its impact on some unknown data. However, the function estimating the global estimating distribution function *EGDF* uses not only parameter *varS* but also a sub-function for automatic decision setting of this parameter by means of testing its impact by using the data. Both subjective and objective decision making find their application in practical analysis.

18.2 Examples

The exhalations of chemical factories are permanently monitored as to contaminations of the environment by dangerous pollutants. Data of Fig. 18.1 are from such monitoring of the exhalations of the organic pollutant PCB28-31. The purpose of this monitoring is to give signals of crossing the "normal" level of the concentration. The problem is in setting the signal level. It must be below the official norm valid for the allowed concentration but over the frequently reached "legal" levels to not generate false alarms. There are three settings of the signal level demonstrated by the Figure:

1. Statistical: Crossing of the mean value plus 3 STD.
2. Statistical: Crossing of the mean value plus 2 STD.
3. Gnostic: Crossing the level of the upper bound USB of the data homogeneity.

All three ways require using of in-line computers and establishing an observing interval—the number of serial measurements. In the statistical cases this number must be large enough to keep it sufficient for a reliable determination of the mean value and of its volatility. In gnostic case—due to its

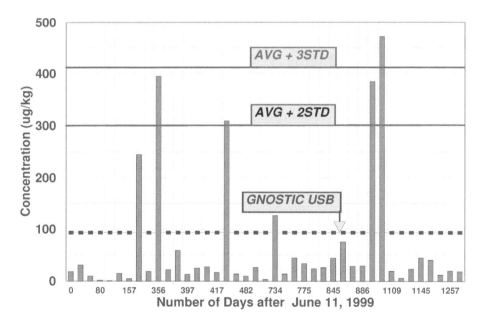

FIGURE 18.1
Monitoring with $varS$ = TRUE

Examples

robustness—it is sufficient to include into the observation interval only several observations sufficient for the gnostic distribution function which estimate the "bounds of normality" (LSB, USB). As the figure shows, the bound USB is crossed in all cases of dangerous exhalations but never in "normal" cases. The data weights remains to be individual according individual uncertainties, but their estimation is made adaptive to changes of actual level of the process.

An objection may arise as to using the STD in statistical case and not a value of probability. To accept this idea it would be necessary to know the current distribution function of the process and this is problematic in statistics, as the process might change and the changes must be respected by the monitoring system. Running the gnostic distribution function in-line provides the system with adaptivity and reliability. The scale parameter S is left to be a constant at each observation interval but it can change from one interval to another with changes in the monitored process.

Another example is connected with one of the methods of monitoring the contamination of rivers by means of catching fishes and analyzing their contamination. Data from such a monitoring were already used above in Fig. 13.1 to demonstrate the forms of the global distribution function and in Fig. 14.6 to show, how the lethal dose of arsenic of fish in rivers can be estimated. The same data can be used to demonstrate the effect of two settings of parameter $varS$ (FALSE and TRUE).

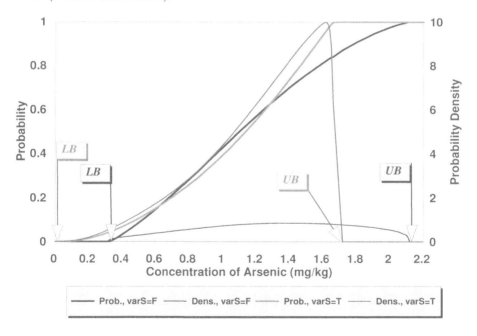

FIGURE 18.2
Comparison of two settings of parameter $varS$

When considering the data as homoscedastic ($varS$ = FALSE), the distribution function is rising much slower than in the case of variable parameter S reaching the upper bound UB of the data support over 2.1 mg/kg which is high over the lethal dose of 1.7 mg/kg determined by using the variable S ($varS$ = TRUE). The conclusion is that the variability of the scale parameter enables the higher flexibility of the global distribution function to be reached.

It is important to remember the gnostic idea "do not violate the data" in connection with the parameter $varS$: each individual data item has the right to be itself, to have its own value and its "personal" variability manifested in the width of its probability density function. Its value is changed in result of an analyst's decision to adapt the distribution function of the sample so to make it closer to Gaussian form by means of transformations. Such an operation is one form of data violation. Applying the same value of the scale parameter to all data of the sample by setting it as a constant is another form of data violation. The best idea is to leave the decision between homoscedasticity and heteroscedasticity on data by checking which version results in a better data fit. The results may be surprising because the difference may manifest itself in such an important feature as homogeneity like in the case of variable *stackloss* which is the fourth column of matrix *stackloss*:

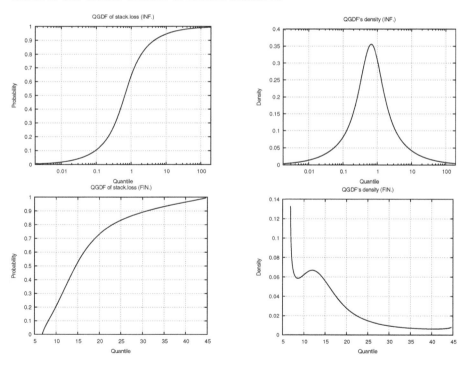

FIGURE 18.3
Distributions of variable *stack.loss* with $varS$ = FALSE

Conclusion

The distribution due to setting $varS = $ TRUE is undoubtedly homogeneous in Fig. 18.3. But the real non-homogeneity is demonstrating with the setting of $varS = $ TRUE in Fig. 18.4.

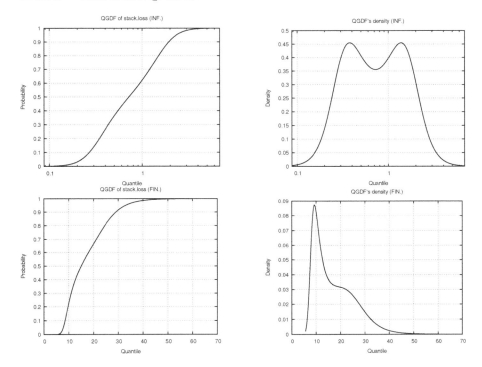

FIGURE 18.4
Distributions of variable *stack.loss* with $varS = $ TRUE

There is an important lesson resulting from this comparison: when testing the data sample's homogeneity, let the data decide on the setting of parameter $varS$.

18.3 Conclusion

The decision making between treatment the data as homoscedastic or heteroscedastic may be done subjectively (manually by the analyst) or automatically (by using the data) by setting parameter $varS$ values FALSE or TRUE. Both ways of making this decision may be useful in data analysis and in applications. The subjective decision may be useful in determination of sensitivity of monitoring systems and in exploration the data features and factors impacting on them. However, the objective decision making may be necessary in testing the data homogeneity.

19
Gnostic Multidimensional Regression Models

Gnostic formulation and solution of robust multidimensional models was published in [54]. The results were implemented as a generalization of the robust statistical method called the IWLS-method (Iterated Weighted Least Squares). The task is defined by a rectangular data matrix of r rows and c columns. One of columns is declared as the *dependent* column while the others as the *explanatory* columns. The task consists in linear case of finding the *best* coefficients c such that using them the linear combination of explanatory columns form the *best* estimate of the dependent column.

A regression function corresponding to the classical definition will be called an *explicit regression*, because it distinguishes between the explanatory variables and one explicitly expressed dependent variable. The IWLS-method (or the method of influence functions) consists of repeated Least-Squares solutions of the set of equations describing the dependence of the model's coefficients on the explanatory vectors:

- The first iteration is the Least-Squares solution of the equations system.

- The residuals (the differences between the modeled and actual values of data) are using as arguments of the influence function to evaluate the weights to be applied to each item of data.

- The next iteration is done as the Least-Squares solution of the weighted equations system till reaching the preset small value of the residuals.

The gnostic notion of explicit regression differs from the statistical formulation in following ways:

- In statistics, data are viewed as random samples taken from a population, the statistical model of which is assumed to be known. A gnostic regression is based on the observed data only.

- Instead of using the usual statistical criteria (e. g. minimum variance, unbiasedness, etc.), gnostic regression procedures minimize or maximize one of several gnostic thermodynamic measures of data uncertainty.

- Classical regression models do not meet the requirements of robustness. The robustness of regression models in robust statistics is achieved by making use some a priori accepted statistical assumptions. In contrast, gnostic regression

DOI: 10.1201/9780429441196-19

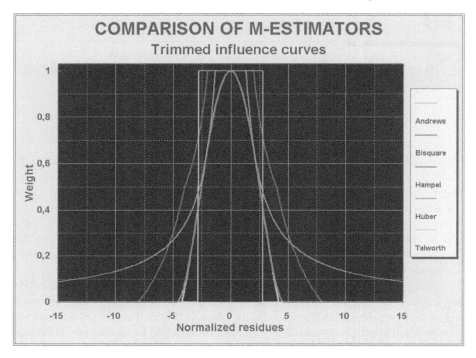

FIGURE 19.1
The statistical trimmed influence functions

models are naturally robust because of their inherent features, which result from the application of the selected gnostic influence function by using the corresponding geometry.

It might be interesting to compare the influence functions used in robust statistics with the gnostic ones. There are two kinds of influence functions used in statistics, the trimmed and smooth. There are sharp edges and limits in trimmed influence functions. Unlike this, the smooth influence functions are differentiable over all data support. The statistical trimmed influence functions are shown in Fig. 19.1.

The disadvantage of the trimmed influence functions is in losing all information on trimmed parts of the distributions function because these parts cannot influence the result.

The gnostic influence functions are smooth, therefore the typical gnostic function is shown along with the smooth statistical ones in Fig. 19.2.

To design the influence function the statisticians use their subjective a priori assumptions on the data uncertainty which may be sometimes useful while otherwise are false. This may lead to bad results in application to some data. The mathematical gnostics is objective in taking all information on the data model from the data.

Formulation of the Robust Regression Problem 207

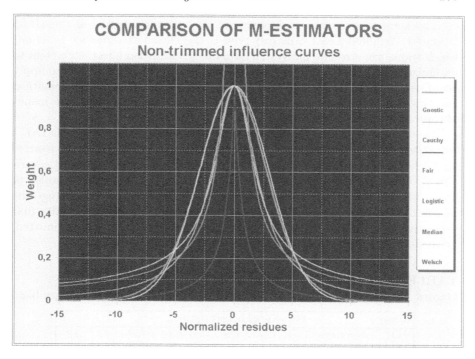

FIGURE 19.2
The comparison of non-trimmed influence functions

19.1 Formulation of the Robust Regression Problem

The formulation and solution of this problem is general enough to include both linear and non-linear versions of cross-section models as well as dynamic models of interrelations between time-series. Explanatory variables may be interpreted as *inputs* and the dependent variable plays the role of the *output* of the model. The model has the form of

$$z = F(c, x) \tag{19.1}$$

where F is a differentiable function of the known type, x the input matrix, z the output variable and c an unknown vector of coefficients, multipliers of the columns of x. It is required to find at least approximately the coefficients of the model.

This approach to the robust regression problem has been originally published in [54]. The paper corresponded to the level of the art of that time and thus needs some upgrades. They will follow along with a short summary of the original paper.

The key problem of the robustness of the regression model of the *IWLS* (Iterated Weighted Least Squares) type is the choice of *influence functions* which transform errors of individual data in each iteration into their weight to minimize their impact on the result. The main idea of the paper was to apply as influence functions the functions of variability which have in mathematical gnostics thermodynamic interpretation. It will be useful to recall the names of these quantities.

There are three pairs of gnostic characteristics interesting as criteria $D(h_c)$ associated with Q- and E-entropy, Q- and E-information, and with the sources of the fields of these quantities (values of which are reciprocal gnostic variances). Criteria are summarized in Tab. 19.1 and their analytic formulae in Tab. 19.2. A numerical multiplier in the definition of D does not change the results of the optimization. The sign of D is chosen to provide a positive first derivative D'_n. Additive constants do not play a role in D, they can be omitted, because only derivatives D' and D'' are needed for the solution.

TABLE 19.1
Gnostic characteristics usable as criterion functions for the regression problem

Case	Gnostic characteristic	$D(h_{c,n})$
Q1	Reciprocal sources of the Q-entropy's field	$h_{q,n}^2/2$
E1	Reciprocal sources of the E-entropy's field	$h_{e,n}^2/2$
Q2	Q-information	$I_{q,n}$
E2	E-information	$I_{e,n}$
Q3	Q-entropy (Q-weight)	$f_{q,n}$
E3	E-entropy (E-weight)	$-f_{e,n}$

Indices c (q or e) distinguish quantification (Q) from the estimation (E) case. All the f_c's and h_c's are for the nth datum, i. e. they should be interpreted as $f_{c,n}$ and $h_{c,n}$ respectively.

TABLE 19.2
Filtering weights and error functions for different gnostic criteria D

Case	Filtering Weight FW	Error Function E
Q1	f_q	h_q
E1	f_e^2	h_e
Q2	1	$f_q \arg\tan(h_q)$
E2	f_e	$f_e \arg\tanh(h_e)$
Q3	$1/\sqrt{f_q}$	$\sqrt{f_q} h_q$
E3	$\sqrt{f_e}$	$\sqrt{f_e} h_e$

Formulation of the Robust Regression Problem

The solution of the problem has the form of an iterative formula of least squares problem

$$c_{k+1} - c_k = \left[\sum_i^n G_i G_i^T\right]^{-1} \left[\sum_i^n G_i E^{-i}\right] \quad (19.2)$$

where

$$G_i = f_i^2 \sqrt{-D_i''} g_i \quad (19.3)$$

with the relative gradient

$$g_i = (F_{i1}', \ldots, F_{im}')/z_{0i} \quad (19.4)$$

and

$$E_i = \frac{sD_i'}{\sqrt{-D_i''}} \quad (19.5)$$

with the scale parameter S. This form of the solution is unusual. The dependence of the system matrix G_i on the data weights by means of influence functions appears always in solutions of this class. However, the multiplier E called *error function* appearing in 19.2 may change the dynamics of the solution significantly. Before showing examples of applications and a large comparison with statistical robust methods, the information on the paper in [54] is to be completed by two tables, especially because the cited paper was only one-sided demonstrating only the estimating versions of the criteria. Leaving out the quantifying ones in the paper was later completed in the unpublished book[1].

The above functions are depicted in Figs. 19.3 through 19.7. The variable on the horizontal axis in Figs. 19.3–19.6 (which show the three categories of Q- and E- weights and errors) is $\Phi_n = \ln(Z_n/\tilde{z}_{0,n})$—the uncertainty measured as the additive residual error (the scale parameter equals 1 in these examples). Fig. 19.7 compares the behavior of the six error functions. To provide a more comprehensive comparison, each of the error graphs also plots the Euclidean error.

To understand properly the behavior of all the filtering functions, it is useful to recall, that by equation 19.4 the filtering weight (FW) amplifies or attenuates the left-hand side of the equation, while the error function E affects the right-hand side. Both of these functions are dependent on the data error (multiplicative residue) $Z_n/z_{0,n}$, but in a different way.

Consider the case of Q-regressions (Q1, Q2 and Q3, in Tab. 19.1), where the filtering weights have the form of Fig. 19.3. Three qualitatively different behaviors of the FW functions are documented:

1. The FW of criterion Q2 is neutral to input uncertainty: the filtering weight is constant like in the case of the classical OLS (Ordinary Least Squares) method.

---
[1] Kovanic P., Guide to Gnostic Analysis of Uncertain Data, Prague, 2018

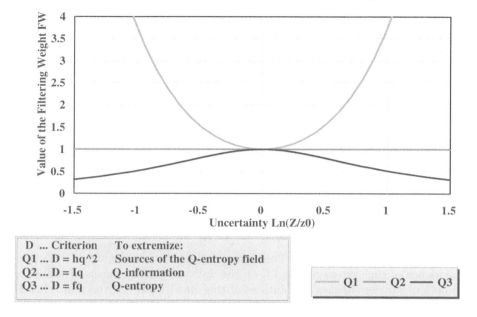

FIGURE 19.3
Filtering weights FW for Q-regression

2. Unlike OLS, a strong outer robustness is manifested in the case of Q1.
3. An inner robustness results in case Q3.

For E-type regressions, all three cases E1, E2 and E3 (Fig. 19.4) lead to the same kind of (inner) robustness but with a different intensity. In all (Q- and E-) cases, the FWs converge to 1, when the multiplicative residue approaches 1 (zero value of the additive residue Φ), which means, that for very weak uncertainty, the differences between it and the OLS-method vanish.

It can be shown, that—for very weak uncertainties—all (Q- and E-) error functions, E, converge to the linear function $2\frac{Z_n - \tilde{z}_{0,n}}{z_{0,n}}$. This can be called Euclidean relative error and it proves, that all six cases of gnostic regression, which were considered, are consistent with the classical regression methodology if uncertainties are sufficiently weak. Moreover, the gnostic regressions were derived without the usual statistical assumptions, they have their own axiomatic justification. This makes them more generally applicable. Moreover, they yield robust results.

The error functions E may be interpreted not only as "influence functions" (as already mentioned) but also as definitions of a Riemannian metric: they determine how an (output) error should be measured. To compare their behavior with that of the Euclidean case, Figs. 19.5–19.7 also show the

Formulation of the Robust Regression Problem

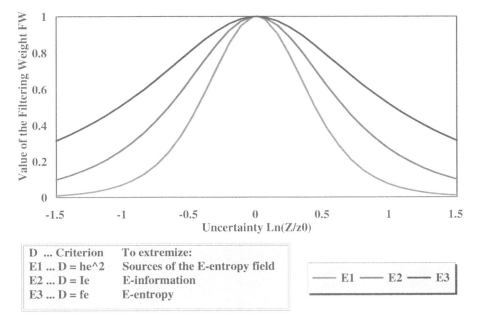

FIGURE 19.4
Filtering weights FW for E-regression

Euclidean metric. For weak uncertainties one can again see the coincidence of the curves with the Euclidean one, however there are large differences, when strong uncertainties are present.

In all three Q-cases, the *E*-function strongly amplifies the effect of uncertainties thus ensuring outer robustness. It is worth noting, that there are three different combinations of input (IR) and output (OR) robustness for Q-regressions:

1. Q1 - IR: outer, OR: mixed,
2. Q2 - IR: neutral, OR: outer,
3. Q3 - IR: inner, OR: strong outer.

The robustness OR for Q1 has been called "mixed," because for some middle intensity of uncertainty (Φ between roughly 0.25 and 0.85) the *E* function rises more slowly than the Euclidean line (more like inner than outer robustness), but for large uncertainties it clearly manifests outer robustness. These differences between the types of input and output robustness in the Q-versions of *D* can be useful in applications, where the makeup and intensities of the input and output disturbances are different.

In contrast, for E-regressions, all input (*FW*, Fig. 19.4) and output (*E*, Fig. 19.6) filtering effects demonstrate an inner robustness, but with

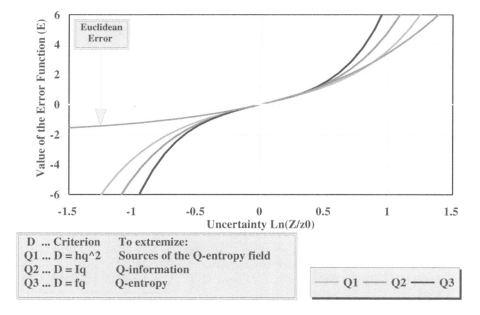

FIGURE 19.5
Error functions E for Q-regression

different intensities. There are other effects to be noted with respect to the error functions in Fig. 19.6.

1. Differences between the E2 and E3 cases are very small, when only the error functions are considered: the type and the effect of output robustness is nearly the same although the criterion functions differ. This, of course, does not mean a complete equivalence, because as shown in Fig. 19.4, the intensity of inner robustness is different.

2. There is a qualitative difference between the error function of E1 and the other two functions: the former behaves in a "saturating" way, while the form of the E2 and E3 error functions is "redescend"[2]. The saturating filter takes all uncertainties beyond a certain limit as "the same," while the redescend filter completely attenuates very large uncertainties.

All six E-functions are shown in Fig. 19.7 plotted against a horizontal axis of multiplicative residues $\frac{Z_n}{z_{0,n}}$ to compare their deviations from those of the Euclidean function (which shows its true linear nature).

---
[2]The term used in robust statistics is used to describe such influence functions.

Formulation of the Robust Regression Problem 213

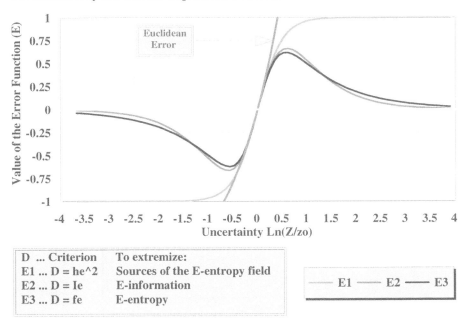

FIGURE 19.6
Error functions E for E-regression

A natural question at this point relates to the choice to be made between the six types and intensities of robustness: "Which one is the best?" Each serves its own purpose and a choice may be alternatively based on

- the prior experience of an analyst with particular types of input and output data,
- the data 'speaking for themselves': that is, to run procedures using all six versions of the gnostic criterion functions sequentially or in parallel and to measure the quality of the results of such a 'pilot' analysis to determine the best D, which can be subsequently applied to analogous situations.

Recall, that all six approaches are optimal, each in its own strictly defined and theoretically justified sense. An important observation should be made as to the geometrical aspects of the measuring errors connected to the regression problems. Euclidean geometry is not curved, it measures errors linearly. However, each of the gnostic error functions shown in Fig. 19.7 represents a non-linear measuring method, which corresponds to a certain Riemannian (curved) geometry. This curvature is what provides robustness to the measurement process: the smaller the local radius of curvature, the stronger the effect of robustness. As Fig. 19.7 shows, the curvature is different at different points along the line. The slope at each of these points is determined separately, by the value of the individual datum. Another factor also influences the

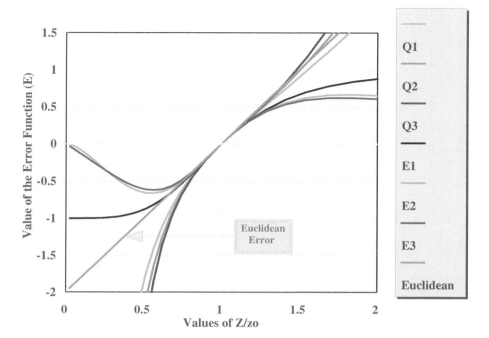

FIGURE 19.7
Comparison of error functions for both Q- and E-regression

curvature at all the data points, the scale parameter (S), the value of which is estimated through optimization of the quality of the model's results. All this again manifests the gnostic credo: "Let the data speak for themselves."

The local curvature is determined by the gnostic functions D, which result from the fundamental features of real data as elements of the commutative group. The metric of uncertain data space is thus determined by the nature of data as mapped real quantities, by "some objectively existing regularities" as specified by Riemann a long time ago.

19.2 Additive and Multiplicative Regression Models

Equation (19.2) has been compared with a classical linear regression model solvable by the least-squares method. However, it could not be taken for a linear relation, because of significantly non-linear weights FW and E dependent on the equation error. Another non-linearity may be caused by the features of the function $F(\underline{C}, \underline{x}_{0,n})$, which is not necessarily linear. The adjective "linear" would not be suitable for the problem solved in previous sections.

However, contributions of individual equation coefficients $C(k)$ to the model's value are additive (with some weights). This type of models can be therefore called *additive regression model*. But it is not difficult to get its multiplicative counterpart by exponentiation of all members of the Equation (19.2). Interactions between contributions of new (strictly positive) explanatory variables become multiplicative. The result is a *multiplicative regression model*. When a need of the multiplicative model of interactions of positive quantities arises, it can be solved after logarithmic transformations of all variables of an additive model. The exponential transformations return then the required multiplicative model.

To compare the two concepts (the statistical and gnostical ones) a large study was undertaken by application of 52 variants of statistical regression models with the gnostic one. It is worth to be described in more detail.

19.3 Comparison of Robust Regression Models

19.3.1 Statistical Methods for Comparison

A rich choice of methods of robust statistics is recently available to immediate applications thanks to the statistical software, e. g. S-PLUS[3] and especially to the rapidly developing R-project of free software environment for statistical computing and graphics [96]. Following classical and robust methods (among others) can be found in these environments:

1. The classical non-robust Ordinary Least Squares Method (*OLS*) implemented by the function *lsfit* (in both of mentioned environments).

2. The class of robust M-estimators (also called "the approach based on influence functions" [95]) is represented by the algorithms applying the Iterated Weighted Least Square (IWLS) method ([95], [26]). There are 10 functions of this type in S-PLUS available differing by the influence functions *wt*: *wt.huber*, *wt.hampel*, *wt.bisquare*, *wt.andrews*, *wt.cauchy*, *wt.fair*, *wt.logistic*, *wt.median*, *wt.talworth*, *wt.welsch*. These methods are applied by the function *rreg*[4]. These will be denoted as *rreg1*, ..., *rreg11*. They are determined by some parameters, the fixed settings of which were reputedly recommended by the method's authors.

---
[3]S-PLUS is a registered trademark of a mathematical and statistical environment recently available with TIBCO Spotfire S+ (http://spotfire.tibco.com/products/s-plus/statistical-analysis-software.aspx)
[4]Cited from the S-PLUS version 6.2.

3. The class of robust estimators implemented by the function *rlm* dependent of following parameters to select:

   (a) The influence function: *psi.huber*, *psi.hampel* and *psi.bisquare*.
   (b) The scale estimator (*scale.est*): "*MAD*," "*Huber*" and "*proposal 2*."
   (c) The method: "*M*," "*MM*."
   (d) The weighting method (*wt.method*): "*inv.var*" and "*case*."

   These 36 methods will be distinguished in the form of *rlm1111*, ..., *rlm3322*, where the numeric indexes denote the selection of parameters.

4. The class of the robust estimators of the Resistant Regressions run by the function *lqs*, which has four versions of the parameter *method*: "*lts*," "*lms*," "*lqs*" and "*S*." They will be denoted by *lqs1*, ..., *lqs4*. All details on functions *rlm* and *lqs* can be found in the environment of the R-project.

There can be thus 52 statistical methods considered for comparison. The last four methods are not exactly comparable with the other 48 methods based on the influence functions[5]. Functions *lqs* apply unitary weights, but satisfy an additional requirement: to ensure a high breakdown point. They are included to evaluate their performance in possible applications, where the breakdown point is not the main problem.

There is a direct link from the 48 considered versions of the Iterated Least Squares Method to the geometric problem. Consider an influence function $W_i = F(e_i)$ where e_i is a fitting error (residual) of the ith equation of a regression model and where W_i is the weight to be given to the equation. Assume at least the local differentiability of the function. Then

$$dW_i = \frac{\partial(F)}{\partial(e_i)} de_i \qquad (19.6)$$

can be viewed as an element of the distance of a path from an initial value to the final value of the variable W_i, measured by using the one-dimensional Riemannian geometry applying the metric tensor $\frac{\partial(F)}{\partial(e_i)}$. The choice of the influence function is thus equivalent to choosing the geometry. Comparison of regression models enables thus 48 geometries for measuring the uncertain data to be compared.

19.3.2 Robust Regression in Mathematical Gnostics

A non-statistical approach to modeling some multivariate relations has been presented in [54]. Similarity to the M-estimators of robust statistics could

---

[5]The *OLS* method can be also included as the case of the trivial influence function of the type $wt.OLS \equiv 1$.

give rise to the opinion that only another version of statistical approach was presented. Indeed, the same Iterated Weighted Least Squares Method can be applied to solve the problem numerically. However, this is the sole common feature, but fundamental differences between the approaches exist as described in [54]:

1. The theoretical fundament is the non-statistical theory of individual uncertain data and small data samples (mathematical gnostics).

2. The only requirement/assumption to the data is their reality in the sense of their being products of measuring or counting procedures satisfying the requirements of consistence formulated by the measurement theory and extended by the gnostic Axiom 1.

3. A kind of Riemannian metric is applied to measuring the errors of the fit (the irrelevance) instead of the additive Euclidean error).

4. This metric is not subjectively assumed but derived from the considered data: "Let the data speak for themselves!"

5. Instead of some purely mathematical criterion functions, uncertainty measures such as data information or entropy are extremized.

6. There are two ways applied to support the robustness:

    (a) Application of weights to both sides of the system equations.
    (b) Filtering of the dependent variable.

    Instead of the ordinary $R^1 \to R^1$ influence function, other functions of the $R^1 \to R^2$ were applied.

7. Iteration of corrections to initial values of model coefficients was used instead of iterations of the coefficients.

8. The output of the method was extended by the results $C_i X_i j$ for all coefficients C_i and for all data $X_i j$ to characterize the shares of data columns on creating the mean result's value.

9. A choice between two kinds of robustness (external and internal) is enabled.

The last feature is to be commented: it was already exposed, that the notion of robustness is ordinarily interpreted as a lower sensitivity to the outliers and to peripheral data of samples. There are two mutually opposite kinds of robustness considered corresponding to different behavior of both (quantifying and estimating) entropies. As an example of the requirement to the increased resistance to internal disturbances of the sample has been the production quality assessment mentioned where the "normal" product quality is a "noise" and the abnormal values represent the emergency "signal." Observations of the cosmos looking for alien civilizations can also be an example.

The standard robustness of a regression model with respect to outliers can be achieved by gnostic criterion functions reviewed in Table 3 of [54]. The opposite robustness can be obtained by using the formulae in Tab. 19.1 and Tab. 19.2 .

Recent implementation of the method used in examples was accomplished by using the environment of the R-project [96].

19.3.3 Data for Comparison

Ten sets of real multidimensional data available in the R-environment have been applied in the comparisons. A complete definition of a dataset has the form of *package::file* in R. A description of a file can be queried in R by the command *?package::file*. The data are shown in Tab. 19.3.

TABLE 19.3
Test data

File	R-package	Size	Dep.	RRE	Mst	GMR
stackloss	datasets	21 × 4	1	0.0100	60.43	59.78
swiss	datasets	47 × 6	1	0.0183	70.14	69.08
airquality	datasets	111 × 4	1	0.3094	42.11	23.44
CPSSW8mM	AER	8505 × 3	1	0.1234	19.92	17.53
CPSSW8fM	AER	6631 × 3	1	0.1181	15.85	13.98
UKconsumption	urca	76 × 3	1	0.0091	6783.36	6722.39
ecb	urca	26 × 5	4	0.0007	101.36	101.29
coleman	robustbase	20 × 6	6	0.0123	2.73	2.70
BD	fields	89 × 5	5	0.1321	29.78	26.60
hachemeister	actuar	29 × 4	4	0.0046	1584.21	1576.57

Data are stored ordinarily as data frames, some of these data include categorical variables, which were to be eliminated to get numeric matrices. Values of RRE are relative residual entropies quantifying the data uncertainty.
Mst is the statistical mean value of the sample and GMR the properly aggregated gnostic mean. Data *stackloss* are called "the guinea pig of multiple regression" [106]. They quantify the dependence of stack loss on air flow, water temperature and acid concentration in a plant for the oxidation of ammonia to nitric acid.
Data *swiss* characterize the dependence of fertility measure for each of 47 French-speaking provinces of Switzerland at about 1888 on five socio-economic indicators (Agriculture, Examination, Education, Catholic, Infant mortality). Data *airquality* quantify the dependence of the ozone concentration in New York (1973) on the solar radiation, wind speed and daily temperature.
Data *CPSSW8mM* and *CPSSW8fM* are subsets of the large set CPSSW8 selected for men and women from the Midwest USA. The earnings of people as a function of education and person's age are demonstrated.
Variables of the data *UKconsumption* characterize the consumers' non-durable expenditure, personal income and consumers' expenditure deflator index, all in 1970 prices.

Macroeconomic data of the Euro zone *ecb* are a time series (1997–2003) of the Gross Domestic Product Deflator, Nominal Gross Domestic Product, Monetary Aggregate M3 and Benchmark Government Bond yield with a maturity of 10 years.

Data *coleman* contains information on 20 schools from the Mid-Atlantic and New England States: staff salaries per pupil, percent of white-collar fathers, socioeconomic status composite deviation, mean teacher's verbal test score, mean mother's educational level and verbal mean test score.

Data *BD* demonstrate the effect of buffers composition of KCl, MgCl2, KPO4, dNTP on the DNA amplification rate.

File *hachemeister* is a data set giving average claim amounts in private passenger bodily injury insurance in 5 U.S. states over 12 quarters between July 1970 and June 1973 and the corresponding number of claims.

19.3.4 Criteria for Evaluation of Methods

Let y be the model's dependent variable vector of the length N and y_f its fit obtained by the model's output. Let w be the vector of resulting weights of the equations obtained by the Iterated Weighted Least Squares method. (All weights w equal to 1 in cases of the Ordinary Least Squares method and of four versions of the *lqs* method.) Denote e as the vector of additive errors of the fit $(y - y_f)$. Let $I(y)$ and $I(y_f)$ be E-information (8.12) of the dependent variable and of the fit. Let $\bar{x}$ be the arithmetic mean of x. Then criteria

$$RobR^2 = 1 - \frac{\sum_i^N (w_i(e_i - \bar{e}))^2}{\sum_j^N (w_j(y_j - \bar{y}))^2} \tag{19.7}$$

(called the Weighted R-square),

$$GMMFE = \exp\left(\sum_i^N (|\log((y_i/y_{fi}))|)/N\right), \tag{19.8}$$

(the Geometric Mean of Multiplicative Fitting Errors) and

$$DivI = \sum_i^N \left(\frac{I(y_i)}{I(y_{fi})}\right)/N \tag{19.9}$$

(the divergence of information of the y and fit) will be computed for each method estimating the model of a particular data set.

The $RobR^2$s value depends on relative distribution of the weights. Like the ordinary R^2, it evaluates the ratio of explained and total variance, but instead of the non-robust sums of squares, it takes into account the weights to increase the robustness. It can thus be used to quantify the quality of the influence function determining the weights.

Both $GMMFE$ and $DivI$ characterize the quality of the fit, the former directly and the latter by comparing the information of the dependent variable and its fit.

The better the model, the higher is its $RobR^2$. The fitting errors increase the $GMMFE$, but the stronger the uncertainty of the fit, the less the $DivI$. To quantify the performance of a model treating a data set, the evaluation respecting all three criteria can be therefore obtained by

$$EvalMet = (RobR^2)/GMMFE \cdot DivI \qquad (19.10)$$

19.3.5 Results of Comparison

Calculation of $EvalMet$ of all 52 methods applied to all 10 data sets revealed the necessity to eliminate the method *psi.median* because of its incomparability with other methods, that make use of more or less all data by application weights less or equal 1. Unlike these functions, the *psi.median* function attaches to a non-zero error e the weight $1/e$, which can far exceed 1. For example, one of weights reaches the value of 460262.3 in the case of the *stackloss* data. The sum of the weights of the other 20 data is only 30.9, thus making their impact on results negligible.

Thus there are only 51 methods to compare in the application to 10 data sets by values of $EvalMet$ forming a matrix (M_{val}) of the size 51 × 10. There are two aspects of viewing this matrix:

1. Horizontally: to see, how well a method works when treating different data sets.
2. Vertically: to compare performance of different methods in application to the same data set.

The geometric mean of values of a M_{val}'s row ($GMEM$) summarizes the performance of a method in application to all 10 sets so that the larger $GMEM$, the better the tested method's version. Vertical means of $GMEM$'s values and of the range of their values corresponding to all versions of *rreg*, *rlm* or *lqs* approach can be used to characterize the diversity of versions. Geometric mean of each of M_{val}'s column (RRE, Relative Residual Entropy) is also interesting as evaluation of the difficulties met in modeling the particular data: the larger RRE, the more difficult the modeling of the data. These values are presented in Tab.19.3 and show that this index takes values over a broad interval from 0.007 to 0.3094 for the considered data sets.

An alternative way of vertical evaluation of the matrix M_{val} consists in the ascending ordering of its columns. Positions of individual methods in application to a particular data set enable us to view the columns as results of methods' competition. The matrix of the positions is denoted M_{pos}.

The results of the comparison done by values of $GMEM$ and by their positions are reviewed in Tab. 19.4. Versions of methods *rreg*, *rlm* and *lqs* are characterized by the minima and maxima of their criterion's values.

TABLE 19.4
Results of comparison of ten best of all 51 methods applied to all the 10 data matrices

Method	$RobR^2$	MeanW	MWErr	GMMFE	GdivergI	EvalMet
Gnostic	0.993	0.710	-4.5e-4	1.115	0.861	0.767
rreg6	0.977	0.662	0.003	1.117	0.864	0.756
rlm1212	0.956	0.905	-0.176	1.136	0.869	0.731
rlm1312	0.956	0.905	-0.176	1.136	0.869	0.731
rlm1211	0.956	0.905	-0.176	1.136	0.869	0.731
rlm1311	0.956	0.905	-0.176	1.136	0.869	0.731
rlm1111	0.968	0.872	-0.241	1.136	0.855	0.729
rlm1112	0.968	0.872	-0.241	1.136	0.855	0.729
rlm3322	0.971	0.849	-0.164	1.135	0.851	0.728
rlm3321	0.971	0.849	-0.164	1.135	0.851	0.728

It may be interesting to see the "top 10" of all of tested methods demonstrated in Tab. 19.4.

19.3.6 Discussion of the Results

The gnostic method appeared to have the best value of $RobR^2$ and it also occupies the best position. The best of the robust statistical method (by both $GMEM$ and position) was the version *rreg6* of the function *rreg* using the influence function *wt.fair*. A more detailed comparison of this method with the gnostic one also confirms the priority of the gnostic method by other statistical criteria. It results in a more significant model.

It can be seen in Tab. 19.4, that not all methods of robust statistics perform better than the classical OLS method at all times. But there is a more serious aspect of their applications, the instability of their performance in dependence on data. They more or less work well with only some data sets. But this is not surprising, because they were developed under some special assumptions on data and to satisfy some special optimization criteria. This is documented by results of the *lqs* function, which ensures a high breakdown point but provides the worst fitting.

Some statisticians oversimplify the interpretation of the Central Limit Theorem when saying to their clients: "Give me more data and the results will be improved." They even perform calculations (design of experiment) of the data amount sufficient for reaching the required accuracy. The Relative Residual Entropy (RRE) was presented in the last column of Tab. 19.3. It shows that the number of observations is not the main problem, because the very large data set $CPSSW8mM$ (8505 × 3) has the third largest RRE (0.1234) from all data sets. This cannot be explained only by a small number of the matrices's rows or by small number of explanatory columns. There are four explanatory columns in the *stackloss* data, but one of them (Acid Concentration) is weakly significant and its contribution to the output is small. In spite of this and the small number of observations (21) the results of treating this small sample are judged according to the difference between *Mst* and *MGR* and are much better then that of the large data set. The key problem is the variability. Take e. g. once more the case of the financial ratios of the IBM from the forgoing chapter. The matrix have only 14 rows. However, the relative variability of the variable $TOTR$ quantified statistically by the ratio $STD(TOTR)/mean(TOTR)$ was 160.02% while for the variable $TLdA$ the same ratio was only 1.64%. The uncertainty RRE was 0.073 in the former and 0.0001 in the latter case. To support the statement on the variability as a main problem in data analysis, the case of the big data sample $CPSSW8mM$ can be remembered. Its first column (*earnings*) has the uncertainty RRE equal to 0.1234 when all 8505 rows are considered. But when considering only 5 rows, the uncertainty is 0.0454. Could we really get a better accuracy by increasing the amount of considered data?

Looking for the explanation of the success of mathematical gnostics in comparison with 50 statistical methods, we return to fundamental problem of the metrics discussed earlier. The OLS method applies the Euclidean geometry and is outperformed by most methods of robust statistics applying the inconstant metrics (influence functions) designed by mathematicians for special classes of data and satisfying some artificial optimization criteria. Unlike this, metrics applied in the case of the gnostic method proposed in [54] are applicable in much more general cases of data. They must satisfy only the requirements to their origin: to be obtained by a quantification process, which is consistent in the sense of measurement theory based on the praxis of measurement. These metrics are determined by the data themselves and can be considered as natural features of the spaces of uncertain real data. The optimization criteria are functions of the natural quantities, entropy and information.

19.4 The Explicit and Implicit Regression Models

Returning to the definition of regression analysis cited at the beginning of the chapter: it was noted, that the separation of one of the variables as the "dependent" one from the others ("explanatory") variables is based on knowledge on the nature of the dependence. There is also another important assumption: that the dependence of the variable to be explained is only one-way. In other words, the "dependent" variable has no influence on the input variables, there is no "feed-back." Regression models of this type can be called *explicit*, because they are based on explicit equations. There are at least three problems with respect to explicit regression models:

The Explicit and Implicit Regression Models

1. In the real world strictly one-way dependencies are the exception rather than the norm. The feed-back is a more frequent case.
2. It is not always possible to solve the model's equations (eg the non-linear ones) with respect to the "dependent" variable, even if such a variable exists.
3. Solutions of over-determined systems of equations—such as explicit regression equations—can be only approximate. This results in inconsistencies if one attempts to exchange the roles of "dependent" and "explanatory" variables.

Indeed, the mathematical definition of a dependence as a function easily introduces a strictly one-way action: "explanatory variables → dependent variable." Modeling real processes is frequently far from easy, because each of the variables being considered is dependent on others. Vivid examples can be found in financial statement analysis: Profitability (measured for e. g. by return on assets, return before tax etc.) is a frequent object of interest to analysts and financial managers interested in discovering, how it depends on other factors such as financial leverage, various turnover relationships, working capital, etc. A regression with profitability PR as the dependent variable is expected to estimate these effects:

$$R_{PR,k} = C_0 + C_1 R_{RWC,k} + C_2 R_{TATO,k} + C_3 R_{FL,k}. \quad (k = 1, \ldots, K) \quad (19.11)$$

Each of the variables $R_{*,k}$ are financial ratios of individual firms: PR is profitability, RWC the relative value of working capital, $TATO$ total asset turnover and FL financial leverage. All of these financial parameters are mutually dependent. When there is a good profit margin, a manager may decide to decrease financial leverage, to improve liquidity, to accelerate the total asset turnover by additional investment (if the demand exists) as well as to take other positive measures, because there are sufficient financial resources. This illustrates the multidimensional "feed-backs," which cause the explanatory ratios to be dependent on the profitability as well as on each other. There are methods in control theory to solve problems with the feed-back, but their application could be even more complex than the initial problem.

The solution of an explicit regression task (19.11) suffers from another serious draw-back: Since the firm's growth is also dependent on the level of the working capital, why not evaluate the dependence of working capital on the other financial parameters using a regression such as 19.11 with working capital as the dependent variable?

$$R_{RWC,k} = c_0 + c_1 R_{PR,k} + c_2 R_{TATO,k} + c_3 R_{FL,k}. \quad (k = 1, \ldots, K) \quad (19.12)$$

Once the parameters have been estimated, the profitability can be expressed as

$$R_{PR,k} = -\frac{c_0}{c_1} + \frac{R_{RWC,k}}{c_1} - \frac{R_{TATO,k}c_2}{c_1} - \frac{R_{FL,k}c_3}{c_1}. \quad (k = 1, \ldots, K) \quad (19.13)$$

The problem here is, that the coefficients $1/c_1$, $-c_0/c_1$, $-c_2/c_1$, and $-c_3/c_1$ are not the same as those of the regression (19.11).

This unpleasant inconsistency is easily explained. A system of regression equations ordinarily contains more equations than unknown coefficients ($K > 3$ in this case) to minimize the uncertainty of the solution. The solution will then depend on using an optimization method and the result is, that the uncertainty of the observed values of the variables in Equations (19.11) and (19.12) is suppressed differently in each regression even if the same optimization method is applied. Obviously, when different results are obtained by the same method from the same data, deciding which of the two solutions is the true one, is difficult.

The conclusion to be drawn here is, that dividing the variables of an explicit regression into categories of "dependent" and "explanatory" is seldom theoretically consistent, because any of them can be either "dependent" or "explanatory" depending on how the problem is set out. Such an analysis introduces an asymmetry into the solution, which leads to inconsistencies in many practical cases.

The desired symmetry in the roles of all the variables can be achieved by using the *implicit form* of a regression:

$$c_1 R_{PR,k} + c_2 R_{RWC,k} + c_3 R_{TATO,k} + c_4 R_{FL,k} = 1 \quad (k = 1, \ldots, K) \quad (19.14)$$

Here, all the variables play the same role. Once a solution for coefficients c_* is obtained, 19.14 can be used to express any desired variable as an explicit function of the others without any danger of inconsistency of coefficients. Further, a single "universal" solution is preferable, rather than a different one for each "dependent" variable. The question of the existence of such a system of equations is answered simply: take the explicit equation system for an arbitrarily chosen dependent variable in the role of the dependent one. This variable V will form the right hand side of the system. Then divide each ith equation by the V_i assuming $\forall i(V_i \neq 0)$.

Using the implicit model, we give to all variables the same "rights" to exercise their impact on the join result. Moreover, the single implicit solution of a task can serve as a source of all explicit equations by simple manipulations with coefficients of the model. And the inconsistencies between individual explicit equations are eliminated.

The problems associated with explicit regressions in economic analyzes are due to the fact, that the inner interactions of the variables of an economic system are nearly as complex as those of a living organism. Indeed, it is impossible

Examples

to state, that any of single parameters (such as e. g. temperature, blood pressure, pulse rate, electric potential, number of blood cells, or the composition of fluids) of a living creature is "only dependent" on other variables and does not also influence the others. The above example using financial statement analysis was linear; however, the same difficulties are also present in the many usual non-linear cases. Problems of the same nature also exist in other application fields. It would seem that some important Laws of Nature are formulated in an implicit form, and that there are similar non-linear interdependencies in economics and in the other social sciences. Returning to the gnostic approach to the regression problem, it is evident, that the discussed formulation can include both explicit and implicit forms. The explicit case was considered in detail, but to undertake an *implicit regression* it is only necessary to substitute a vector of "1's" (denoted by $\underline{1}$) instead of the observed 'output' values, $z_{0,n}$:

$$\underline{x}\underline{c} = \underline{1}. \tag{19.15}$$

Underlining is used in this equation system to emphasize, that $\underline{x}$ is a matrix (of all variables) and both $\underline{c}$ and $\underline{1}$ are vectors. Calculations of an implicit regression can require modifications of algorithms ordinarily used in variance analysis, because they assume a non-zero variance of the "dependent" variable. We have already seen, that an implicit model (19.15) exists if all dependent data of the explicit model have non-zero values. However, the *intercept* cannot be used in an implicit model, because it would make the system singular.

It is not difficult to adapt the function designed for explicit regression model for implicit mode. It requires substituting the whole data matrix with a parameter (Dep) saying which of column numbers should denote the dependent variable. With $Dep = 0$ the implicit model will be computed while with positive Dep the explicit one will be created. The gnostic regression function is called *GWLS* (Gnostic Weighted Least Squares). It has many parameters determined by the data, but the main parameter is the data sample and the second is Dep determining if the model is implicit ($Dep = 0$) or explicit with the Dep'th column of data as the dependent variable ($Dep > 0$). It has been noted that when running the implicit modeling, the constant term called *Intercept* could not be included because of causing the singularity of the system. However, in cases of explicit models it plays an important role. Its non-zero value indicates the insufficiency of the choice of explanatory variables for finding all impacts on the dependent variable: there exist other variables influencing the dependent variable which are not included into the model.

19.5 Examples

One of important gains of multidimensional models is their role as a training tool for the control of a multidimensional object. Take the example of driving a car. To control the direction, you have a rule, to control the speed there is a gas pedal and a gear and two kinds of breaks. Learning of driving takes a while and requires a praxis in the street in spite of the fact that the effect of individual control means is obvious and known. However, take the praxis of controlling a business. The example of financial ratios of the IBM includes eight variables the impact of which on the business's behavior is not as simple as in driving a car. And e. g. COMPUSTAT Data Item List includes 349 financial variables of a business which could be used to create a mass of financial ratios for modeling. It is obvious that each real model must miss some variables and that estimation of the errors of missing variables is worth of the effort. To demonstrate this and the role of the implicit model, three models of the financial statement analysis will be shown by using the historical financial statement data of the IBM:

1. $M1$ ... the explicit model of the total return of the $FINRATIBM$ matrix ($Dep = 8$) including the constant term (int) called *Intercept*,

2. $M2$ ... the explicit model of the total return of the same matrix ($Dep = 8$) without the constant term ($int = FALSE$),

3. $M3$ ... the implicit model of all 8 variables ($int = FALSE$) explaining the 'dependent' vector consisting of numbers 1 ($Dep = 0$).

The model $M1$ has the following structure:

TABLE 19.5
Explicit model $GWLS < -(FINRATIBM, Dep = 8, int = T)$ including the *Intercept*

Var.	C	EC	$\bar{x}$	Ex	Cx	ECx	pC	pCx
Intcpt	8.894	0.378	1.000	0.000	8.894	0.000	0.000	0.000
CLdA	-10.230	0.402	0.415	0.021	-4.241	0.210	0.000	0.000
TATO	0.629	0.399	0.255	0.023	0.160	0.015	0.139	0.000
RWC	-13.425	0.169	0.077	0.018	-1.033	0.247	0.000	0.001
EBITdA	59.058	0.984	0.030	0.007	1.800	0.442	0.000	0.001
TLdA	-5.011	0.454	0.759	0.013	-3.802	0.065	0.000	0.000
PS	-0.011	0.0005	75.225	33.225	-0.817	0.361	0.000	0.041
P/E	-56.798	0.981	0.013	0.005	-0.764	0.293	0.000	0.022

The parameter Dep determines that model should be explicit with the dependent variable No.8 ($TOTR$) while parameter int requires inclusion of the *Intercept*. Following items form the model:

- Var. ... the variable's name,

- C ... the value of coefficients,

Examples 227

- EC ... standard error of the k's coefficient (k is the matrice's row number),
- $\bar{x}$ ... mean value of the k's column of the data matrix,
- Ex ... standard error of the k's column,
- $\bar{C}x$ ... mean of the data's k's column multiplied by the coefficient,
- ECx ... standard error of the data's k's column multiplied by the coefficient,
- pC ... probability of the erroneous accepting the insignificant variable,
- pCx ... probability of the erroneous accepting the insignificant variable multiplied by the coefficient.

The application of statistical notions of 'mean', 'standard error', 'significance' and 'p.vals' may seem strange but the procedure is applicable not only for data requiring the robust handling but for all data. Application of statistical parameters obtained in robust cases is taken as conditional approximation.

Some important messages are obtained by column ($\bar{C}x[,k]$) showing how are individual variables sharing the mean value of the dependent variable. The sum of shares is low equaling only 0.197. The variables *Intercept*, *TATO* and *EBITdA* increase the dependent variable *TOTR*, but other variables and especially the *CLdA* and *TLdA* compensate the large part of the positive impact. The items of the column deserve a comment: a large impact of the constant term *Intercept* is stronger than two other positive impacts together and an extreme negative impact of both current *CLdA* and total debt *TLdA* represent a warning for the financial manager: use the debts only when really necessary. The relatively strong positive impact of *TATO* and the negative impact of the *RWC* represent a recommendation on what the manager should concentrate its attention. The high impact of *Intercept* represents a warning for the analysts: the unexplained impact is not negligible, it would be worth to trying other variables.

A noteworthy information can be obtained by squares of values *STD(Cx[,k])* the sum of which makes 1.633. The square share of the *PS* and *P/E* on this sum equals to 40.0%. These variables contribute to the *TOTR*'s uncertainty value in the strongest way. This may be explained by the incapability of the market to estimate the price of the shares of *IBM* in an adequate way.

To show the effect of ignoring the constant term a comparison of the most important columns of both models $M1$ and $M2$ is presented in Tab. 19.6. The "dependent" variable is *TOTR* in both models.

It is obvious from this comparison that inclusion of the constant term should not be missing in explicit models.

The implicit model of the same data is in Tab. 19.7.

The coefficients of the implicit model substantially differ from that of the explicit model, but it is caused by necessity to multiply them by values of x which are of different magnitude. The shares on the "dependent" variable have

TABLE 19.6
Comparison of explicit models of the data FINRATIBM including ($M1$) and missing ($M2$) the Intercept

Variable	C M1	$\bar{C}x$M1	C M2	$\bar{C}x$M2
Intcpt	8.894	8.894	0	0
CLdA	-10.230	-4.241	-10.293	-4.267
TATO	0.629	0.160	-9.138	-2.326
RWC	-13.425	-1.033	-6.312	-0.486
EBITdA	59.058	1.800	32.993	1.006
TLdA	-5.011	-3.802	7.112	5.395
PS	-0.011	-0.817	0.003	0.193
P/E	-56.798	-0.764	45.313	0.610

TABLE 19.7
Implicit model of the data FINRATIBM

Var.	Cs	EC	$\bar{x}$	Ex	$\bar{C}x$	ECx	pC	pCX
CLdA	0.640	0.061	0.415	0.021	0.266	1.31e-02	0.000	0.000
TATO	-0.375	0.049	0.254	0.023	-0.095	8.69e-03	0.000	0.000
RWC	0.416	0.071	0.077	0.018	0.032	7.67e-03	0.000	0.001
EBITdA	-1.232	0.225	0.030	0.007	-0.038	9.23e-03	0.000	0.001
TLdA	1.027	0.046	0.759	0.013	0.779	1.33e-02	0.000	0.000
PS	0.0001	0.000	75.225	33.225	0.011	4.76e-03	0.013	0.041
P/E	3.446	0.237	0.013	0.005	0.046	1.78e-02	0.000	0.022
TOTR	-0.0004	0.002	0.116	0.187	-0.000	8.41e-05	0.850	0.549

different signs with respect to those of the explicit model, but the dominant role of both debts *TLdA* and *CLdA* in control of the business economics is again confirmed.

There also are other results of regression analysis: the model is applied to the data to obtain the modeled data and their weights and errors. To compare the explicit and implicit versions, they are shown all in Tab. 19.8:

It is necessary to take into account the different scales of both model when considering their modeled values *E-mod* and *I-mod*, weights *E-ws* and *I-ws* and errors *E-Res* and *I-Res*. The explicit model E- is for the absolute value of the *TOTR* expressed in dollars while the implicit model I- uses the dimensionless number 1.

There also are summaries of both models presented in Tab. 19.9. They show the means of columns of the important models's characteristics, their robust R-squares, mean data weights, weighted errors and geometric mean of weighed errors.

TABLE 19.8
Explicit (E) and implicit I model of data FINRATIBM

Y. q.	E-mod	E-ws	E-Res	I-mod	I-ws	I-Res
96 4	0.229	0.979	-0.010	0.994	0.010	5.9e-03
97 1	-0.093	1.000	0.002	1.005	1.1e-30	-4.8e-03
97 2	-0.018	5.6e-04	0.336	1.000	1.000	7.1e-05
97 3	0.170	0.999	0.007	0.984	1.2e-35	1.6e-02
97 4	0.262	5.5e-24	-0.273	1.000	1.000	3.1e-05
98 1	0.030	0.826	-0.035	0.993	0.004	6.9e-03
98 2	0.087	0.965	0.020	1.000	0.852	-5.6e-04
98 3	0.108	0.987	0.013	1.000	0.987	2.3e-04
98 4	0.125	1.62e-03	0.312	1.000	0.994	-6.7e-05
99 1	-0.024	0.963	-0.014	1.004	8.2e-05	-4.2e-03
99 2	0.463	0.995	-0.003	1.000	1.000	2.1e-05
99 3	-0.059	0.993	-0.004	0.992	9.5e-04	8.4e-03
99 4	-0.113	0.997	0.005	0.989	4.8e-05	1.1e-02
00 1	-0.323	2.7e-26	0.418	1.000	1.000	1.8e-05

TABLE 19.9
Summary of explicit and implicit models of data FINRATIBM

Model	$RobR^2$	MeanW	MWErr	GMMErr
Explicit	0.994	0.693	-8.53e-05	2.234
Implicit	0.873	0.489	-8.09e-07	1.004

19.6 Homogeneity of an MD-Model

The homogeneity of a data sample is his fundamental feature because it provides the data with a special identity. It manifests itself by tendency to be close together, to make a cluster and to separate itself from other data. We have seen this feature at homogenous probability distributions of global type for which the test of homogeneity consisted in checking of just one maximum of the data density function of the convex type. Application of this test can be extended for the multidimensional models: The MD-model is homogeneous when all his variables are homogeneous. The multidimensionality of an MD-model is thus testable as the repeated uni-dimensionality. Two ways were considered for homogenization of a non-homogeneous data sample, the homogenization by elimination and by weights. There is a third way usable for MD-data: subsequent eliminating the individual data appearing over the perimeter of the cluster. This way is laborious but in a cooperation with a robust regression model it works as shown in Fig. 19.8: there is a comparison of the gnostic regression model with 11 known statistical regression models.

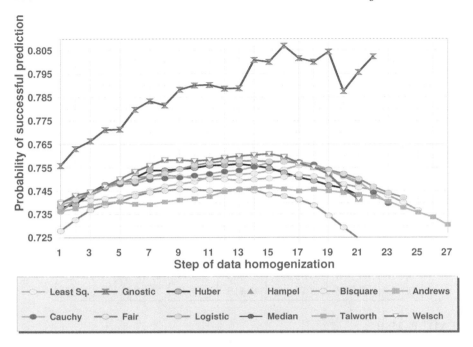

FIGURE 19.8
Comparison of robustness of regression models

Unlike the statistical models reaching a maximum and then falling, the gnostic model with increasing number of homogenizing steps exhibits a steady rise.

19.7 An Important Multidimensional Model

An information on the large survey comparing loading of inhabitants with the dangerous pollutants living close to a large chemical plant ("a dirty place") with contamination of people living far from an industry ("a clean place") was already presented in connection with distribution functions. The survey was using both objective and subjective methods:

Objective: Measuring the contamination by POPS (Persistent Organic Pollutants) of the fat in blood of samples taken from all participants of the survey.

Subjective: Collecting personal data from questionnaires filled by participants relating to their way of life, their health state and habits.

Application of a non-traditional indicator ("the accumulation of POPS," the

An Important Multidimensional Model

contamination divided by the age of the participant) was also mentioned. All the data obtained in this manner were then used to identify two multidimensional models (for the "dirty" and "clean" places) for evaluation of impact of individual variables on the accumulation of POPS. The results are summarized in Fig. 19.9.

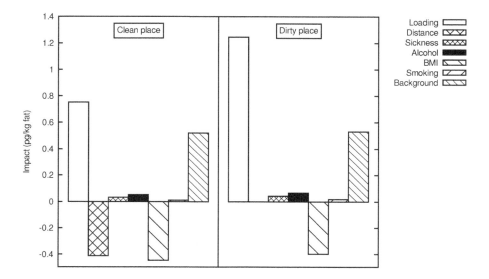

FIGURE 19.9
Impacts on factors on accumulation of POPS in "dirty" and "clean" places

There were following variables included:

The dependent—measured and calculated: The mean personal accumulation of POPS and the intercept, the constant term of the model.

The explanatory—by questionnaires: The distance between the "clean" and "dirty" places, sickness measured by number of diagnosis, drinking alcohol by amount, body-mass index, smoking by amount of cigarettes.

Some interesting conclusions can be drawn from the graph:

1. The idea that a change of living place would enable an escape of dangerous pollutants is bad. Actually only decrease of about 63% cents of the contamination can be expected.
2. The habits of smoking and drinking alcohol as well as the sickness and their impacts increase the danger of accumulation of POPS only slightly and are independent on the living place.

3. The fat people are relatively less contaminated because the same amount of POPS is "diluted" in a larger amount of fat. This effect is also independent on the place of life.

4. The shocking result is the coincidence of intercepts of both models proving that a substantial danger of contamination exists independently on the living place, at least in the middle Europe over which appears the known continental maximum of bad air causing the acid air and adding to smog. The local transport also adds to contamination independently on the living place.

The bad impact of the local transport on the concentration of the lead and cadmium in soils and on the health of children has been proved in Polish-Czech study[6]. An example of a high contamination by cadmium is in Fig.19.10

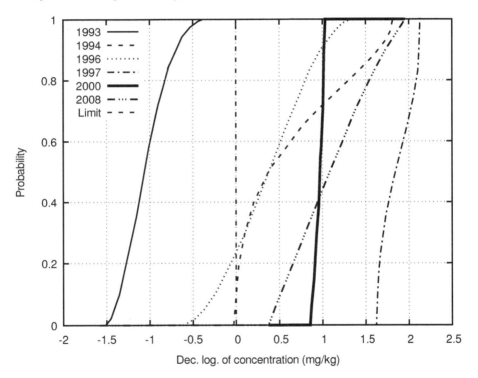

FIGURE 19.10
Impacts of Pb and Cd on hemoglobin of children

---

[6]Kovanic P., Impact of the Pb and Cd on hemoglobin (The Institute of Public Health, Ostrava (2010))

19.8 Applications of the Robust Regression Models

There exists a broad scale of applications of the regression models:

1. To explore the interdependence of columns of a matrix object.
2. To control a variable column.
3. To quantify impacts of individual variables.
4. To serve as a robust filter.
5. To enable multidimensional cluster analysis.
6. To help in testing of homogeneity of a multidimensional object.
7. To enable homogenization of a multidimensional object.
8. To order the multidimensional matrices/models.
9. To robustly estimate trends and dynamics of a process.
10. To robustly estimate the auto- and cross-correlations.
11. To visualize the matrices.
12. To robustly estimate relations of a general type.
13. To reveal and quantify the critical points of processes.

Some of these tasks have been already exposed: Task No.1 is solved by values of the model coefficients, task No.2 by using the explicit model computed directly, or—better—by recalculating the implicit model. The solution of the task No.3 is available from outputs of the modeling. This and other applications will be considered in detail below.

19.9 Conclusions

There exists a large number of statistical robust regression models, mostly based on influence functions. There also exist several versions of the gnostic influence functions which minimize or maximize variables defined in gnostic theory. The difference between statistical and gnostic functions lies in the metric used in quantification of uncertainty. Statistical influence functions and their modifications are designed without willingly revealing their relations to geometry. Unlike this, gnostic influence functions respect the geometry of the quantifying and estimating errors. A large numerical study comparing 50 versions of statistical robust regression models with the gnostic ones has shown in applications to ten cases of real data the superiority of gnostic influence functions over the statistical ones.

The implicit alternative of the standard explicit regression models extends the applicability of the regression modeling and offers some advantages.

20
Data Filtering

20.1 Filtering

The notion of filtering is a very general one including techniques of removing undesirable components from a mixture of materials. In the application to uncertain data, it means a more or less complete separation of the true value from its observed value contaminated by the uncertainty. The simplest filter in statistics is represented by the separation of the mean value as the best estimate of a constant true value while minimizing the standard deviation evaluating the uncertain component of the observed value. It has been shown, that the statistical operator $mean()$ is applicable only to uncertain data with a weak uncertain component. In more general cases the aggregation law should be applied which is additive to non-linear functions of observed data and to four gnostic data characteristics. More general aggregation operator $Gmean()$ thus enters the game generating the mean value by one of three manners described in chapter "Measuring the data uncertainty" as MGQ, MGR or MGE. The idea of improving the value of an individual data item by giving to it a weight dependent on its individual error is implemented by different ways:

- by application of an influence function within the robust regression modeling,
- by using the weights applied in proper aggregation operation,
- by application of an implicit model.

20.2 Total Data Variability and Its Components

The standard statistical model represents the uncertain data as an (additive) composition of the true value and an uncertain component. Many statistical methods take the true data values as unknown constants. In previous chapters this book did not consider the nature of data variability in detail and

DOI: 10.1201/9780429441196-20 235

created the model of total variability when introducing the Ideal Gnostic Cycle and its thermodynamic parameters. Really, all data changes are quantified by parameters f_J and f_I not regarded their origin. As such, they are rather measure of total variability of data than measures of uncertainty, the changes of which only contribute to the total variability masking the actual changes of true data values. A multidimensional object's state or process is described by an MD-matrix composed of observed data. A regression model of such a matrix can specify not only state and interactions of all variables, but also all changes of variables independently on their origin. The coefficients of such a model can explain the total variability of data. An attractive task appears, to develop a method of matrix filtering suitable to transform a real matrix into a *clean matrix* the regression model of which would ignore the uncertain data component describing only the impact and interactions of the actual data and their actual variability. To do this needs introduction of the *implicit model* as a variant of the regression model.

20.3 Filtering by Regression

The idea of using the multidimensional regression model for filtering of data is based on the fact that the model reveals and quantifies the interdependencies between different variables. Therefore the estimate of the dependent variable prepared by using the explanatory variables may be less uncertain than the dependent variable estimate itself due to information transferred from the explanatory variables: the explanatory variables "remember" the behavior of the dependent variable thanks to the information passed by interdependencies. This can be demonstrated by an example.

Let us compare the total variability of vector *stack.Loss* which is column number 4 of matrix *stackloss* with the variability of its estimate is obtained using the multidimensional explicit regression model (Tab. 20.1).

The statistical evaluation of STD confirms the improvement of the vector quality obtained by modeling.

The implicit model of estimated vector has a much smaller total variability characterized by Tab. 20.2.

TABLE 20.1
The total variability of original vector *stack.Loss* and of its estimate by the regression model

Vector	Mean	STD
Original	17.52	10.17
Estimated	17.17	8.74

TABLE 20.2
Comparison of total variability of explicit and implicit models of vector *stack.loss*

Model	Mean	STD
Explicit	17.17	8.74
Implicit	1.14	0.66

The advantage of the implicit modeling grounds in possibility to be applied not only to a matrix, but also to a vector like in the case of vector *stack.loss*. This feature along with a low variability enables thinking on the causes of the filtration effect. It has been shown in the study on the regression method [54], that the disturbing impact of uncertainty on data can be compensated by weights applied to each data item reciprocal to item's estimated error. Such weights are used in both explicit and implicit regression models.

It is useful to look for such weights in application of the proper data aggregation.

20.4 Filtering Effect of Proper Data Aggregation

The proper data aggregation is applied additively not to data values but to their gnostic characteristics which are hyperbolic and/or trigonometric cosines and sines of data items' values. Each data item is a sum of the true value and of the
uncertain component causing its error. Take for instance the cosine. The ratio of its value and of the sum of all aggregated cosines is an individual weight given to the particular data item. The result of aggregation depends on the applied geometry. For the statistical mean the standard error of aggregated vector *stack.loss* is 10.17162, while for the quantification case (hyperbolic cosines, minimization of f_J) is 2.256 and for estimating case (trigonometric cosines, maximization of f_I) we obtain 1.968. However, when considering the case of the geometry R for which properly aggregated vector Gmean(.)[[2]][3] is obtained by $\sqrt{error(filtered\ quantification) \times error(filtered\ estimation)}$, the standard error will be 0.157. It means that using the weights corresponding to the phases of the Ideal Gnostic Cycle, we are meeting a significant smoothing effect. This way of filtering can be called *thermodynamic* as the extremes of thermodynamic variables f_J and f_I are found.

20.5 Improving the Matrix Quality

There are several motivations for exploring the filtering effects on vectors and matrices:

1. The already cited Linnik's proof of statistical statement on improving the data by introducing the weights equal to reciprocal variance of the sub-sample.
2. The idea on the estimation error as individual "fault" of a data item.
3. The gnostic improvement of the regression method [54].
4. The positive features of implicit regression model.
5. The smoothing effect of the proper data aggregation.
6. Three ways of data aggregating.
7. The advantages of the implicit model over the explicit one.

Matrices are very useful instruments in solving practical problems. Results of their application are degraded by data uncertainty. Good filtering of matrices is thus an important problem. Let us consider the applications of the thermodynamic filtering to improve the matrices. There is an important requirement to such a filtering: the relations between columns of the filtered matrix characterized by the multidimensional model should be sufficiently approximated by the MD-model of the filtered matrix.

Let *GWLS(x, Col)* be again the gnostic version of the weighted least-squares model where Col is the column-number of the dependent vector. (In case of the implicit model *Col=0*). Let $y = GWLS(x, 0)[[2]][, 1]$ be the result of the implicit model. Then the improved/filtered (*clean* matrix or vector) Fx is obtained as

$$cx = x/y \qquad (20.1)$$

If x is a vector of the length L, then the result will be a number repeated L-times.

The design of the filtered matrix is shown in application to the already considered 21×4 matrix $M = stackloss$. Its filtered value FM is obtained by substitution of M's columns by their thermodynamically filtered values. If V_i is the i-th column of M, then its filtered value FV_i is calculated as $Gmean(stackloss[, i])[[2]][, 3]$. If the MD-model of the matrix M is $MD(M)$ and of the filtered matrix is $MD(FM)$, then the errors of the approximation of the coefficients of the model matrix expressed in percent are in Tab. 20.3.

When judging on the errors, it would be useful to consider the variability of the variables, they are shown in Tab. 20.4.

TABLE 20.3
Errors of coefficients of the model of filtered matrix stackloss in percent

Variable	Coeffs	Std(C[k])	Mean(x[,k])	STD(x[,k])
Air.Flow	-4.249	-0.365	0.010	0.096
Water.Temp.	12.719	0.606	0.005	0.058
Acid.Conc.	1.797	0.206	0.001	-0.012
stack.loss	-5.960	-4.523	1.486	3.543

TABLE 20.4
Variability of original and termodynamically filtered matrix *stackloss*

Variable	Matrix	f_J	f_I	RRE
Air.Flow	original	1.0101	0.9902	0.0100
Water.Temp.	original	1.0102	0.9899	0.0101
Acid.Conc.	original	1.0020	0.9980	0.0020
stack.loss	original	1.1374	0.8960	0.1187
Air.Flow	filtered	1.0101	0.9902	0.0099
Water.Temp.	filtered	1.0102	0.9900	0.0101
Acid.Conc.	filtered	1.0020	0.9980	0.0020
stack.loss	filtered	1.1374	0.8960	0.1187

Variabilities of the filtered vectors are equal to that of the original vectors, because the vector representing the uncertainty has constant items. The variability of the stack.loss is much larger than that of other variables. This may explain the largest error of this variable.

It may be interesting to have a look at the thermodynamical filtering of the financial matrix $FINRATIBM$. The least uncertain column of this matrix is the $FINRATIBM[,5]$ ($TLdA$), the largest is $FINRATIBM[,8]$ ($TOTR$) as shown in Tab. 20.5.

TABLE 20.5
Extreme variabilities of original and thermodynamically filtered matrix $FINRATIBM$

Variable	f_J	f_I	RRE
TLdA	1.0001	0.9999	1e-04
TOTR	1.0793	0.9325	0.073

The corresponding extreme errors of the coefficients of MD-models expressed in percentage are 0.3595 for the variable $TLdA$ and 3.516 for the $TOTR$. Repeating of thermodynamic filtering does not give an improvement of results.

TABLE 20.6
Separation of the uncertain component of the data variability

Matrix	Variable	fJ	fI	RRE
stackloss	Air.Flow	1.0101	0.9902	0.0100
clean	Air.Flow	1.0000	1.0000	1.0000
uncertainty	Air.Flow	0.0012	0.0011	0.0435
stackloss	Water.Temp.	1.0102	0.9899	0.0101
clean	Water.Temp.	1	1.0000	0.0000
uncertainty	Water.Temp.	0.0102	0.0101	0.0049
stackloss	Acid.Conc.	1.0020	0.9980	0.0020
clean	Acid.Conc.	1.0000	1.0000	0.0000
uncertainty	Acid.Conc.	0.0020	0.0020	0.0000
stackloss	stack.loss	1.1374	0.8960	0.1187
clean	stack.loss	1.0000	1.0000	0.0000
uncertainty	stack.loss	0.1374	0.1040	0.1384

20.6 Cleaning of Matrices

The advantage of the implicit model is that the impact of data variability onto the matrix can be checked by controlling the errors of satisfying the requirement to keep all the "dependent" values equal to 1. Using the actual values determined by the implicit model as weights of the input data creates values of a *cleaned matrix* which will be denoted CM. Regression model run for the CM gives an exact implicit model satisfying practically exactly the requirements of all "dependent" values to equal 1. Let us demonstrate the effects of matrix cleaning by examples. Function $GWLS$ is an implementation of the Weighted Least Squares method where the parameter Dep determines the column number of the matrix M of the dependent variable. Substituting $Dep = 0$ generates the implicit model. Consider the operation

$$X = GWLS(M, 0) \tag{20.2}$$

result $X[[1]][,1]$ of which is the vector of coefficients of the model while $X[[2]][,1]$ contains the model results the required value of which is 1 for all variables. Apply these values as weights of the input data M:

$$CM = X/X[[2]][,1] \tag{20.3}$$

to get the clean matrix. Really, when running the operation

$$Y = GWLS(CM, 0) \tag{20.4}$$

for the input matrix $M = stackloss$ we obtain the resulting vector $Y[[2]][,1]$ the maximum absolute deviation of which components from 1 is $1.3323 \cdot 10^{-15}$.

Cleaning of Matrices

Matrix CM is a clean matrix. Moreover, comparing the model coefficients of both matrices M and CM we find

$$X[[1]][,1] = Y[[1]][,1] \ . \tag{20.5}$$

Two different matrices have the same model. It can be seen, that further repeating the operation of matrix cleaning does not bring a change. However, the results of application of a clean matrix are more precise than that of the uncleaned one. This may play a significant role on practice. Consider the matrix $FINRATIBM$. Its results of the implicit model deviate from 1 with the standard error $STD = 0.00836$ while for its clean matrix this number reduces to $1.95 \cdot 10^{-15}$. Operation of matrix cleaning can be applied to vectors, as well, like to medical data lh. Instead of the error $STD = 0.219$ the cleaned vector has the error of the implicit model $STD = 6.85 \cdot 10^{-17}$.

TABLE 20.7
Explanation of removing the uncertain component of the data variability

No.	$Res1$	$Res2$
1	0.9774	1
2	1.0217	1
3	0.9719	1
4	0.9522	1
5	1.0206	1
6	1.0363	1
7	1.0757	1
8	1.0657	1
9	1.0426	1
10	0.9348	1
11	0.9853	1
12	0.974	1
13	0.976	1
14	1.0434	1
15	0.9978	1
16	0.991	1
17	0.9182	1
18	0.9574	1
19	0.9688	1
20	0.9556	1
21	1.089	1

The advantage of the matrix cleaning consists not only in substantial lowering of the errors of the implicit model, but also in the complete separation of the disturbing ("random") uncertainty (suppressed by the cleaning) from the actual variability of the data which is controlled (explained) by the model of the clean matrix.

The effect of separation of the uncertainty deserves an explanation. The clean matrix has another feature not seen on the first sight. Denote the (joint) model coefficients of the original and clean matrix by CFM and try the following equations:

$$stackloss \tag{20.6}$$

$$CM(stackloss) \tag{20.7}$$

to get the results:

All results obtained by the clean matrix equal to 1 while the original matrix produces variables. This features may be useful especially for predictions.

20.7 Conclusions

A matrix of variable data may be improved by filtering which lowers the impact of uncertainty of data. The filtering operation may be implemented both by regression model and by using the thermodynamic filtering. The most efficient filtering effect can be achieved by using the advantages of implicit regression model offering the cleaning of matrices. The operation of matrix (and vector) cleaning separates the actual uncertainty from controlled/explained variability of the true data by eliminating the impact of uncertain data components.

21
Decision Making in Mathematical Gnostics

There are at least three Czech proverbs relating to one of the most important problems of life. They sound approximately in the following way:

    Don't overdo it![1]

        Too much of a good thing.[2]

            Everything in moderation![3]

The last of the proverbs is a double of the English saying "within a measure." All of these expressions attract your attention to a point where a good thing turns to become something undesirable, bad, unpleasant, impractical, unsuitable or even dangerous. Let us call these points *critical*. Such an important point must be determined quantitatively to become an actual regulatory instrument for decision making. Such a decision making is usually carried out under the impact of the uncertainty. As such, it happened to be an important problem of statistics. It relates to many practical tasks, for instant some of them:

1. Testing of distributions of probability of uncertain events.
2. Testing of the domain bounds of an uncertain function.
3. Testing of hypotheses on uncertain events.
4. Identification of outlying events.
5. Decision making in market operations (buy, hold, sell).
6. Quality control of production.
7. Classification of uncertain events.
8. Testing the homogeneity of samples.
9. Comparison of laboratories or products.

It is not within scope of this book to describe the statistical solutions of these problems in detail. It is sufficient to mention the main characteristics of the statistical approach: the deciding points are determined as chosen quantiles of a probability distribution. There are two points in this approach calling for criticism:

---
[1] Nic se nemá přehánět (in Czech)
[2] Všeho moc škodí (in Czech)
[3] Všeho s mírou (in Czech)

1. The subjectivity of the choice of the distribution.
2. The subjectivity of the choice of decisive quantile, of bounds of the tolerance interval or of the significance level of the test.

The problem of the distribution is that only rarely the "named" statistical distributions (*NSD*) exist on practice precisely in their defined form. Even the distributions based on some regularities of physics like e.g. the Poisson distribution can falsify the experiment. It can describe the atomic decay of a material only approximately because it is very difficult in practice to observe one single isotope without presence of another one with a different decay constant. Even the application of the most desirable and advantageous normal distributions (ruling Taleb's Mediocristan) is subjected to criticism because of doubts if the theoretical conditions of its existence are fulfilled in the given particular case. The choice of the adequate distribution from the large amount of *NSDs* can be problematic as well because the Nature makes the actual distribution mostly different from their "pure" theoretically expected form. Moreover, the attempt to make a real distribution one of the *NSDs* (e.g. Gaussian) can result in falsifying the parameters of the distribution.

The choice of the decisive quantile is another source of subjectivity because this choice can pre-determine the result of the decision making. There is a rule in statistics to choose the level of significance before the analysis of data. However, the praxis of strict requirement to support the scientific conclusions by results of statistical tests led to a "technique" to choose the level of significance of statistical statements only after the data are available and to adjust the significance level to ensure the positiveness of the test.

Such a statistical praxis can be called liberal or even democratic.

21.1 Datacratic Decision Making in Mathematical Gnostics

There is one wise principle known for a long time in statistics: "Let data speak for themselves!" This principle motivated the aim of mathematical gnostics to develop especially the design of probability distributions not based on a priori assumptions but on real data values. However, to know what data say about their distribution is not sufficient for an objective solving the decision problem. Therefore, the aim was directed to extension of the "consulting" role of the data by their participation in decision making by establishing the decisive points. Finding and using the decisive points determined objectively by the data can be called *datacracy*. Most of such points usable as decisive appeared already in previous chapters in connection with explanations of the thermodynamic physics of data uncertainty and development of the gnostic data analysis. Along with technique of getting these instruments their datacratic applications were discussed:

1. The decision making: additive versus multiplicative data.

2. The finality of the domain of real data with the bounds LB and UB of the data support and their roles in distribution functions.

3. Probability and probability density values along with the contribution to information of each data item obtained by global and local distribution functions and by thermodynamic filtering.

4. The homogeneity and homogenization problems of distribution functions, the decision homogeneous/non-homogeneous data sample.

5. The scale parameter S and its role in distributions, the decision making homoscedastic/heteroscedastic data samples.

6. The decision making on metrics 'Q'/'E' (quantifying/estimating).

7. The application of the bound UBS as the critical point of the environmental control (18.1)

8. The bounds LB, LSB, ZL, $Z0L$, $Z0$, $Z0U$, ZU, USB and UB of interval analysis enabling the distribution of elements of a data sample into 10 classes.

9. Bounds LCB, CRE and UCB of a series of sub-samples in calibrating analysis of a sample.

10. Parameters of the thermodynamic analysis of a sample: GMQ, $STDq$, GMR, $STDr$, GME, $STDe$, f_J, f_I and RRE. These parameters are obtained in estimation of the properly aggregated mean value and characterize sample's uncertainty.

11. The testing of hypotheses by means of distribution functions (10.9, 10.8).

12. Comparison of laboratories, producers and market chances (to be discussed below).

13. Critical points in analysis of general relations (to be discussed below).

Further there is also a datacratic role, the choice of functions to be maximized and minimized in analysis: instead of some formal mathematic functions, the thermodynamic characteristics f_J are minimized and f_I maximized to optimize the Ideal Gnostic Cycle.

21.2 Conclusions

Decision making under uncertainty is subjective by using probability distribution functions resulting from arbitrary a priori choice of the analyst and decisive points as bounds of tolerance intervals also based on its subjective point of view. Unlike this, all decisions are based in mathematical gnostics on the thermodynamic features of uncertainty and on using sample's data to estimate the natural critical points. This makes the decision making objective, independent on the free will of the analyst but unique and natural. All results of gnostic analysis are determined by the data. The analysis can be thus called datacratic, because data values decide everything.

22
Comparisons

Comparing of objects is a frequently asked task of data analysis. Some should simply decide yes/no like in answering a question of acceptance of an object from the point of view of a norm or of concordance of parameters of objects originated by different methods or made by different producers. A decision based on evaluation of variants is frequently required like, e. g., in ordering of investment chances: which object is better from a given point of view. Comparisons must satisfy all requirements to data analysis like objectivity, robustness, satisfactory precision, sensitivity and reliability, all measurable under the uncertainty of results.

22.1 Comparisons of Measurement of Toxicity

One of important tasks of environmental research is measurement of toxicity of many materials both originated in Nature and resulting from human activity:

1. Contamination of air,
2. contamination of waters,
3. contamination of food,
4. contamination of soils, and
5. others.

Surprisingly large amount of toxic effects were found in different natural materials such as lead and cadmium and many others and also in pollutants obtained when doing ordinary activity like burning. However, more dangerous materials appear due to the progress of technology, like persistent organic pollutants (*POPs*). A lot of methods of the toxicity measurement has been developed and their efficiency studied. An example of comparisons of methods of toxicity measurement cited from the results of the Zdravotní Ústav Ostrava (The Institute of Public Health, Ostrava) can help in understanding of the size of the problem (22.1).

A series of measurements of a fixed number of organic pollutants was done on water samples of Czech and Moravian rivers and subjected to

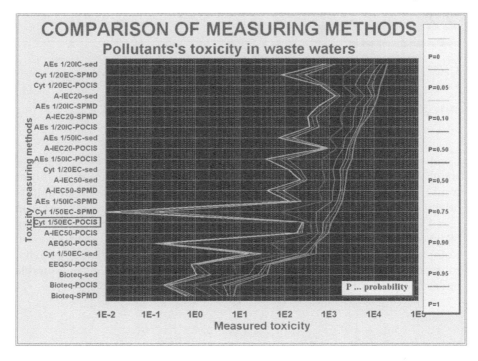

FIGURE 22.1
Comparison of methods of measurement of toxicity of POPs

23 methods of toxicity measurements. Gnostic global distribution function was then calculated for each of methods and quantiles of ten probabilities $(LB, 0.1, \ldots, 0.9, UB)$ was depicted in the Figure 22.1. Probability ranges $([LB, UB])$ appeared to occupy a broad range of more than six orders of magnitude. As the most sensitive method appeared the method Cyt1/50EC-SPMD[1] the ratio of UB/LB exceeded 1000, the least sensitive was Cyt1/50EC-POCIS with the ratio about 10/1.5.

The sensitivity and range of results are not sufficient for decision making as several measurement methods should be applied. Consider a simple example of decision making between four most popular measurement methods of toxicity of river waters: Bentos, Dreissena, Plants, and SPMD. Six requirements to achieve better quality results were formulated:

1. external consistency, (comparison with other measurements),
2. internal consistency, (comparison within these measurements),
3. informativeness, (sample's quality of information),
4. precision, (measured/mean value),

---

[1](Semipermeable Membrane Device)

5. homogeneity, (by local distribution functions),
6. relative sensitivity (by error/(observed value)

and methods for evaluation of these features quantitatively prepared. Results are presented in Fig. 22.2.

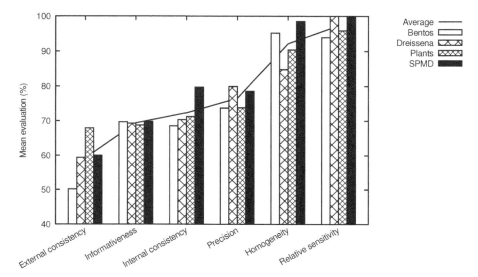

FIGURE 22.2
Comparison of methods of measurement of the contamination by POPs

Comparison of measurements done on the same materials in different laboratories is another important task. Several methods based on the mathematical gnostics applicable to this purpose were already considered above. There is an official norm on comparison of laboratories *ČSN EN ISO/IEC 17043*. It could be interesting to show an application of the method recommended by this norm with application of the gnostic methods.

22.2 Comparison of Measurement of Concentration of Cannabinoids

According to the Wikipedia, in 2014 there were an estimated 182.5 million cannabis users (3.8% of the population aged from 15 to 64)[2]. Frequency of taking the cannabis is documented by increasing interest of young population and by frequent changes of legislative limiting or allowing cultivation

---
[2]Narcotic Drugs 2014, International Narcotic Control Board 2015, p.21

and distribution of cannabis and its derivatives. Measurement of concentration of cannabinoids, the principal psychoactive constituents like *THC* (δ-9-tetrahydrocannabinol) and *CBD* (cannabidiol) deserves a rising interest of researches and producers. Thanks to Dr. Tomáš Ocelka, CEO of the EHSS Co. Prague, data obtained from the comparisons of measurements of cannabinoids measured by two laboratories were made available for publication. The laboratory No.1 was of the EHSS Co., No.2 was the laboratory of the Prague University of Chemical Technology. Basic comparison was performed several times according to the norm *ČSN EN ISO/IEC 17043* by a specialized agency and resulted in numbers *En* with a conclusion, that all measurements of two laboratories were in agreement. It has been then decided to apply gnostic method for the comparison of standardized method with the novel approach. The data are in Tab. 22.1.

TABLE 22.1
Statistical evaluation of measurements of two laboratories

Cannabinoid	L1a	L1b	L2a	L2b	En
1CBD	0.1700	0.0800	0.2700	0.1000	-0.89
1CBDA	0.6900	0.2100	0.9000	0.2500	-0.64
1THC	0.0140	0.0040	0.0200	0.0080	-0.67
1THCA	1.0450	0.0140	0.0380	0.0150	0.34
2CBD	0.0450	0.0090	0.0500	0.0150	-0.29
2CBDA	0.1010	0.0200	0.1600	0.0468	-1.16
2THC	0.0081	0.0016	0.0065	0.0020	0.62
2THCA	0.0078	0.0016	0.0120	0.0078	-0.53
3CBD	0.0210	0.0063	0.0200	0.0040	0.13
3CBDA	0.0730	0.0219	0.0440	0.0090	1.22
3THC	0.0035	0.0011	0.0036	0.0007	-0.08
3THCA	0.0078	0.0023	0.0073	0.0015	0.18

The official protocols on comparison performed according to the norm *ČSN EN ISO/IEC 17043* suffered by several imperfections:

1. The general conclusion of the performance (all measurements of two laboratories were in agreement) appeared to be in contradiction with the norm which requires for the number *En* compliance of the condition $|En| < 1$. The value *En* was in cases of *2CBDA* and *3CBDA* -1.16 and 1.22.

2. An estimate of the true value of each pair of estimates was missing in the protocols along with statements on values of errors of 48 measurements.

3. Although the protocols are called "Protocol on two-side comparison," a statement which laboratory was better is missing.

4. A statement on measurement precision of each laboratory is missing.
5. A comparison of measurements uncertainty was not done.
6. Precision of measurements on different cannabinoids differ but their comparison is missing.
7. Conclusion of comparison in language of probability or information is missing.

If we limit ourself on the question of the agreement of laboratories like the official evaluation, it is sufficient to play a little with the data and look up at the result by the ordinary eyes.

22.2.1 Comparison of Multiplicative Errors

By comparing two positive errors e_1 and e_2 by using the multiplicative way one determines their ratio $\frac{e_1}{e_2}$. There is an advantage of the multiplicative errors: it is not necessary to estimate the mean value of the errors, the ratio's value 1 corresponds to agreement of the errors. Let us consider a matrix RMC consisting of three columns. The first column is formed by ratios of errors $L1a/L2a$, the second column of $L1b/L2b$ and the third of ratios $\frac{L1a/L1b}{L2a/L2b}$. In the case of an exact agreement of laboratories, all the ratios L1a/L2a, L1b/L2b and L1aL2b/L2aL1b would equal 1. Tab. 22.2 shows that this is not the case of agreement.

TABLE 22.2
Multiplicative disagreements of measurements of two laboratories

Cannabinoid	L1a/L2a	L1b/L2b	L1aL2b/L2aL1b
1CBD	0.630	0.800	0.787
1CBDA	0.767	0.840	0.913
1THC	0.700	0.500	1.400
1THCA	1.184	0.933	1.269
2CBD	0.900	0.600	1.500
2CBDA	0.631	0.427	1.477
2THC	1.246	0.800	1.558
2THCA	0.650	0.205	3.169
3CBD	1.050	1.575	0.667
3CBDA	1.659	2.433	0.682
3THC	0.972	1.571	0.619
3THCA	1.068	1.533	0.670

To consider the results of two laboratories as an agreement would be equivalent to the statement "all values of the interval equal 1."

The official comparison can be therefore evaluated as unsatisfactory and the norm as outdated.

22.3 Requirements to the Advanced Comparison

Results of measurements of laboratories to be compared must satisfy several requirements:

1. Homogeneity of the data sample.
2. Precision.
3. Reliability.
4. Informativeness.
5. Measurability.
6. Comparability.

Comparison of laboratories is based on comparisons of measurements performed on the series of the same samples in two or more laboratories. It is required to estimate a qualitatively delimited quantity of some objects. The only chance to verify the identity of the objects by each of the laboratories is by analysis of results. The first (and categorical) requirement is the homogeneity of each sample. The homogeneity manifests itself by grouping of several measured values into a cluster, which has just one maximum of the global gnostic probability density distribution function. Another maximum appears by presence of an outlier or of a sub-sample of data of a nature different from the homogeneous sub-sample. When it is required to measure the variable A, then neither an appearance of a variable B nor the result of A measured with an error closing it to B is not acceptable.

Estimation of results' precision of each laboratory needs the estimates of the true values of each measurement. This can be done by evaluation of results of all laboratories.

The reliability of results is dependent on their uncertainty. The estimate of the uncertainty of the results is a must of the analysis and is an important characteristic of a laboratory accompanying its precision.

Informativeness of each measurement is measured by the amount of contribution to information yielded by the measurements.

Comparability is proved by comparisons with other laboratories. The exact concordance is not to be expected, but the differences are to be acceptable from the point of view of praxis.

22.4 Preparing Data for Analysis

There are two 12 data values obtained from measurements of both laboratories available to be compared. Each measured item of the table was measured in both laboratories, therefore the universal concordance of measurements would be achieved if all equations

$$L1a = L1b \tag{22.1}$$
$$L2a = L2b \tag{22.2}$$
$$L1a/L1b = L2a/L2b \tag{22.3}$$

hold, otherwise an additive error appeared. The true value of difference of both side of equations in all the equations would be zero. However, to judge on priority of a laboratory, an estimate of the true value of each equation side is necessary. The statistical mean of the difference ($\text{mean}(eL - eR)$) of the equation sides could not help, because the critical point would divide the difference into two equal parts. Therefore the gnostic mean Mgn obtained by the true aggregation method was applied. Results are in Tab. 22.3 called as matrix ME:

TABLE 22.3
Error estimates of measurements of two laboratories

Id	ErL1a	ErL2a	ErL1b	ErL2b	ErL1a/b	ErL2a/b
1CBD	-0.0443	0.0557	-0.0095	0.0105	-0.2703	0.3047
1CBDA	-0.0980	0.1120	-0.0191	0.0209	-0.1536	0.1607
1THC	-0.0027	0.0033	-0.0017	0.0023	0.5420	-0.4580
1THCA	0.0036	-0.0034	-0.0005	0.0005	0.3607	-0.3203
2CBD	-0.0024	0.0026	-0.0026	0.0034	0.9175	-0.7492
2CBDA	-0.0261	0.0329	-0.0106	0.0162	0.8949	-0.7363
2THC	0.0008	-0.0008	-0.0002	0.0002	1.0063	-0.8062
2THCA	-0.0019	0.0023	-0.0019	0.0043	2.1364	-1.2001
3CBD	0.0005	-0.0005	0.0013	-0.0010	-0.7492	0.9175
3CBDA	0.0163	-0.0127	0.0079	-0.0050	-0.7036	0.8520
3THC	0.0000	0.0001	0.0002	-0.0002	-0.8634	1.0977
3THCA	0.0003	-0.0002	-0.6713	-0.0004	-0.6713	0.8040

It is convenient for the analysis to form a 36×2 matrix TOL with the first column made by concatenation of the results of the first laboratory ($TOL[,1] = c(ME[,1], ME[,3], ME[,5])$) and the other with errors of the second laboratory ($TOL[,2] = c(ME[,2], ME[,4], ME[,6])$). These tables are ready for analysis.

22.5 Analysis of Measurement Errors

22.5.1 Characterization of Data

The data are additive and the suitable metric for analysis is found to be estimating ('E'). The data can be considered as heteroscedastic by using a variable scale parameter. They are non-homogeneous—a bad mark for the measurement technology.

22.5.2 Comparison by Parameters

The first step was to estimate the uncertainty (*diversity* of the choice of the values of samples to be measured by the values of the relative residual entropy RRE along with estimates of mean errors (bias) and uncertainty of all the 72 values of the columns of Tab. 22.2. Then the mean values and uncertainty were estimated for each of the laboratories. Results are in Tab. 22.4.

TABLE 22.4
Parameters of error estimates of measurements of two laboratories

Lab.	Mst	STDx	GMR	STDr	RRE	min(Err)	max(Err)
Both	0.0329	0.4892	0.0274	0.0049	0.0246	-1.2001	2.1364
L1	0.0627	0.5309	0.0517	0.0108	0.0282	-0.8634	2.1364
L2	0.0030	0.4492	0.0024	6.00E-04	0.0738	-1.2001	1.0977

The summary of statistical errors of both laboratories was 0.0329 ± 0.4892. It means that the relative error was 6.725%, much higher than at ordinary technical measurements. It can be caused by outliers as shown in Tab. 22.4 by columns min(Err) and max(Err). The range of errors of laboratory 1 is broader than that of laboratory 2. From this point of view laboratory 2 is better. However, the relative entropy RRE of the laboratory 1 is much lower than that of laboratory 2. It means, that in spite of larger outliers, the measurements of laboratory 1 are less uncertain. This conclusion is not supported by the statistical standard errors.

Let us now compare the mean values of errors. Both laboratories measure by biased ways, mean values are positive instead of zero-valued. Both statistical means *Mst* and gnostic *Mgn* ones show the bias of the laboratory 1 larger (0.0627 and 0.0517) than that of the laboratory 2 (0.003 and 0.0024). This may be caused by the non-homogeneity of the samples and must be analyzed by distribution functions. Taking the measurement as a "sportive match" of two laboratories the winner of which was that with the less value of the absolute error, one can evaluate the *sportive score* 13:21 in favor of the L2 with 2 draws of equal errors.

Analysis of Measurement Errors

Mean robust (gnostic) mean values of GMR weighted by probabilities were 0.7954 for L1 and 1.1519 for L2. The later result was better because of being closer to the true value 1.

Values of the uncertainty (RRE = the relative residual entropy) are measures of the uncertainty of each laboratory. They were 0.0282 for L1 and 0.0738 for L2. From this point of view was the laboratory L1 more reliable then L2.

Probability analysis of multiplicative errors of all cannabinoids has shown, that probability of better results was 0.527 in laboratory L1 and 0.473 in L2.

Important results of the gnostic analysis includes estimation of robustly filtered values of each individual result of measurements, density of its probability and its contribution to overall information brought by results. These are important for evaluation of different factors on which each measurement depends. Comparison of contribution of each laboratory to the overall information can be done by sums of product of the probability of the measurement with its contribution to information. Results are in Tab. 22.5 for all columns of the input data:

TABLE 22.5
Mean informativeness of measurements of two laboratories

Variable	L1a	L1b	L2a	L2b	L1a/L1b	L2a/L2b
Informativeness	0.1240	0.1280	0.1250	0.1243	0.0867	0.0895

Values of mean informativeness of individual variables differs only slightly, but those of laboratory 2 were better twice and that of laboratory 1 only once.

The relative residual entropy RRE allows the evaluation of measurability to be introduced as $1 - RRE$ and applied to individual cannabinoids ordered in Tab. 22.6 by decreasing measurability:

TABLE 22.6
Measurability of cannabinoids

Cannabinoid	Add. measurability	Mult. measurability
CBD	0.9846	0.9666
THC	0.9813	0.9513
CBDA	0.9578	0.9354
THCA	0.9453	0.8729

These differences in measurability should be respected by the designer of the measurement, e. g. by the numbers of samples to be measured.

22.6 Conclusions

Results of comparison of two laboratories performed according to the norm *ČSN EN ISO/IEC 17043* have been found as unsatisfactory and the norm is out-of-date. Unlike this, the gnostic method resulted in quantified answers to questions on priority of each of the compared laboratories related to measuring both additive and multiplicative data errors, to estimate probability of their values and of the success of laboratories in the comparison. These results were accompanied with quantification of uncertainty of the design of the experiment, of each measurement and of the measurability of individual tested cannabinoids.

23
Advanced Production Quality Control

23.1 Exploratory Analysis

The modern society tends to substitute the work of the people by using automatically controlled robots. It is a recent development of the century's old idea "the donkey-work to machines, thinking to people." However, there also is a donkey-work of mental activity like computing. The primary task for computers was to make people free off non-creative mental work like calculations. Computers took over the functions of calculations from data analysts by providing them with specific functions but unfortunately not completely. The practical implementation of statistics suffers from some high requirements as to thinking. Take for e. g. exploratory analysis required as the obligatory first phase of data analysis introduced in[1] the following way:

> The classical methods of statistical inference depend heavily on the assumption that your data is outlier-free and nearly normal, and that your data is serially uncorrelated. *Exploratory Data Analysis* (EDA) uses graphical displays to help you obtain an understanding of whether or not such assumptions hold. Thus you should always carry out some graphical exploratory data analysis (EDA) to answer following questions:
>
> 1. Do the data come from a nearly normal distribution?
> 2. Do the data contain outliers?
> 3. If the data were collected over time, is there any evidence of serial correlations between successive values of the data?

A text-book of statistics [33] sees the task of exploratory data analysis in "revealing the specific features of the data and of their suitability for the subsequent statistical treatment." The bookrequests a series of operations on data and tests requiring subjective analyst's decision making. The assumption on normality of data distribution is non-existent in mathematical gnostics, the gnostic analysis including the distribution function is applied to data of arbitrary distributions. Moreover, computing of gnostic distribution functions

---
[1] S-PLUS 4 Guide to Statistics, Data Analysis Products Division, MathSoft, Seattle (1997)

includes the basic cluster analysis of data (the lower outliers, not outlying data, the upper outliers). There also are further algorithms enabling to analyze the data feature, to classify them as to their relation with their best estimate. The subjective phase of analysis EDA is thus not necessary and its tasks can be entrusted to the objective automatic data analysis.

23.2 Automation of the Exploratory Analysis

The idea of "Let the data speak (and decide) for themselves" implemented in gnostic procedures makes it possible to automatize the exploratory analysis by computing only the global distribution function. The necessary input information includes only a data sample. If some other information on data is available, it is used in the form of function's parameters, but if these parameters are missing, they are automatically estimated in the analytical process. The function's output consists of the following:

1. A list of the input data sample, of their a priori weight, of their censoring, of their a posteriori weight, of their filtered value, of their probability and probability density, of their relation to the central cluster (signed by '0') or to the cluster of the lower (-1) or upper ($+1$) outliers, the data value in the standard region, the item's individual scale parameters and the data item's a posteriori weights.

2. A list of information on the data obtained in analysis: Data additivity ($FALSE$ or $TRUE$), the type of metric (Q, R or E, the constant or variable scale parameters, the bounds of the data support (LB and UB), evaluation of data uncertainty, the location of maximum density, the information quality of the sample, the mean and maximum horizontal and vertical fitting errors of the probability, the kind of the distribution function (global G or local L), parameter $varS$ ($FALSE$ or $TRUE$) deciding between constant and variable scale parameter, sample's homogeneity ($FALSE$ or $TRUE$), the mean trend of sample's data, the error which would be caused by neglecting the curvature of the data space (in %), the data autocorrelation, the mean of distribution-perpendicular fitting errors, the robust sample's mean value and its residual entropy.

3. A list of numerical presentations of distribution functions and densities over the original (finite) and infinite data support.

4. A figure containing four graphs of probability distributions and densities over the finite and infinite data support.

All information necessary for subsequent data analysis is thus obtained in this way without a priori guessing on the data features. If the result says that the function is not homogeneous, it depends on the analyst whether there are reasons for the homogenization.

The full power of the gnostic approach is achieved in combination of different kinds of distribution functions as was shown by the examples of data NIST12 and NIST37. After revealing the non-homogeneity, local distribution functions help to "dismount" the data into homogeneous sub-samples.

23.3 On the Necessity of Data Inspection

Unfortunately, the elimination of exploratory data analysis does not bring the definite end of the necessity of human participation in preparing the automatic analysis. The phase called *data inspection* is necessary to be passed before entrusting the analysis to computers. The task of this phase is to make sure that the data to be analyzed satisfy the high recent requirements. As known from https://www.go-fair.org/fair-principles/, in 2016, the FAIR Guiding Principles for scientific data management and stewardship were published in *Scientific Data* to improve the Findability, Accessibility, Interoperability, and Reuse (FAIR) of digital assets. The principles of the process of FAIRification includes requirements to data which should be described with rich metadata, provided with globally unique identifiers, license and registration and made computer-available. These principles are recently only recommended but the technological progress will surely make them legally enforced. The data inspection should ensure a maximum care of reaching the FAIR level along with more elemental requirements to data completeness for the goals of analysis.

23.4 Data Certification

The high level of automation and the objectivity of the analysis getting all information from data and not from some subjective assumptions allows to issue quality certificates for features of data samples automatically to summarize sample's parameters and to formulate warrants quantitatively. Data certificate is prepared by an automatic program. It is implemented by application of the gnostic interval analysis which makes the use of hidden features of data samples to sort them with respect to 11 intervals of values and to provide them with probability of the sorting.

According to the Wikipedia, eight dimensions of product quality management can be used at a strategic level to analyze quality characteristics. The concept was defined by David A. Garvin. The dimensions include:

1. Performance, especially involving measurable attributes.
2. Features—additional characteristics.
3. Reliability—the likelihood that a product will not fail within a specific time period.
4. Conformance—Conformance is the precision with which the product or service meets the specified standards.
5. Durability—the length of a product's life.
6. Serviceability—the speed with which the product can be put into service when it breaks down.
7. Aesthetics—the subjective dimension, which represents the individual's personal preference.
8. Perceived Quality—the quality attributed to a good or service based on indirect measures.

All these dimensions can cause the necessity of decision making. They may have some quantitative aspects and as such they may involve measurements and data treatment. It is useful to cite the statement 2.3.6.1 of norm ISO 9000:

> Decisions based on the analysis and evaluation of data and information are more likely to produce desired results.

This is because of existence of risk, which is according to the statement 3.7.9 of this norm an effect of uncertainty—the state, even partial, of deficiency of information related to understanding or knowledge of an event, of its consequence, or likelihood. To implement the principles of mathematical gnostics, the gnostic requirements of production control should be respected including into the data analysis following elements:

Data inspection including data inventory and preparation of resulting plan of necessary actions to provide the missing data, estimation of them and of censored data,

Homogeneity test by using kernel estimates of distributions and possible data homogenization,

Data classification by means of interval analysis and probability estimates of memberships in intervals,

Data modeling to describe bi- and multidimensional both deterministic and uncertain relations between variables and their critical points,

Uncertainty definitions quantitatively describing the impact on results of analysis.

An example of such analysis follows.

23.5 Example of Advanced Quality Control

It was decided to apply gnostic methods of advanced production quality control to data collected by the E&H Services (Prague) in connection with the IPSIC project (International Passive Sampling Interlaboratory Comparison). This enabled to demonstrate the requirements and methods of gnostic data analysis to results obtained by eight laboratories by measurement and treatment of eight pairs of measurement of the same samples of waters contaminated by 16 kinds of POPs (Permanent Organic Pollutants). Participated laboratories were identified by numbers: 3, 4, 6, 7, 8, 9, 12 and 13. The individual POPs are coded in the following way:

TABLE 23.1
Short coding of individual organic pollutants

Obj1	Naphthalene
Obj2	Acenaphthylene
Obj3	Acenaphtene
Obj4	Fluorene
Obj5	Phenantrene
Obj6	Anthracene
Obj7	Fluoranthene
Obj8	Pyrene
Obj9	Benzo(a)anthracene
Obj10	Chrysene
Obj11	Benzo[b]fluoranthene
Obj12	Benzo[k]fluoranthene
Obj13	Benzo[a]pyrene
Obj14	Benzo[g,h,i]perylene
Obj15	Dibenzo[a,h]anthracene
Obj16	Indeno[1,2,3-c,d]pyrene

Two samples of the same contaminated water were prepared by the organizer of the comparison (E&H Services, Ostrava and Prague, Dr. Tomáš Ocelka) and sent to foreign laboratories.

The severe requirements of the FAIR level could not be applied because they were not yet obligatory and not all provisions for using them were available. However there is sufficiently enough elementary features of data substantial for an inspection. A rough and preliminary evaluation of quality of results of individual laboratories can be obtained by an inspection of the basic data features—their quantity and their evaluation by the laboratories. Such an inspection is summarized in Tab. 23.2 in the form of table of bad points (penalties). They were obtained in the following way: *Completeness:* The case of NA (not available) instead of the number is evaluated as the incapacity of the laboratory to measure the sample's value. The number of samples is 16, the penalty for one NA is 1/16. *Censoring:* To declare the result of measurement for left-censored means that the laboratory is incapable to quantify such a low amount of the material. The penalty is again 1/16 for each left-censored data item. *Suspect:* The pairs of samples A and B were prepared experimentally. It is practically impossible to prepare two samples exactly equivalent each to other. The measuring is also not precise. Therefore to declare the same "measured" value for results A and B gives rise to suspect, that either the measurement was too rough to register the difference (a bad case) or there was a try to artificially eliminate the measuring error (also a bad case). Therefore appearance of equality of A and B should be evaluated by the penalty of 1/8. A couple of strange results appeared in results of laboratory 7: Data item of Phenantrene is declared as the same of Fluorathene equaling 1148.9125. However, the robust means of all laboratories were 442.54 for Phenantren and 814.97 for Fluorathen. Therefore the additional penalty of 1/8 for laboratory 7. *Numerics:* Some laboratories are using only integers to demonstrate their measured results, although in some cases the differences/additive errors might show the difference expressed in values of 0.1. This is interpreted as a fault and subjected to a penalty. *Uncertainty:* Precision of measurement can be evaluated as measurement uncertainty. Measuring errors of two quantities A and B is evaluated in statistics by using the formula |A - B| (the additive error) or A/B (the relative error). The relative error is also obtained by using the additive operation, but applied to logarithms (the geometric mean). The additive operation is in both cases applied to data values. However, as shown in MG, this is true only in application to precise data. According to the second axiom of MG, the proper aggregation law of uncertain data is different. Therefore the measuring errors of values A and B and the measurement uncertainty are to be evaluated by means of gnostic function Gmean. Effect of this function generally depends on the amount of uncertainty. In case of samples of length 2 and only in this case (the case of the mean of A and B) the proper aggregation gives the same mean value as the geometric mean. Evaluation of means of A and B must be done before the start of the data analysis implemented by homogeneity tests. These tests are necessary for identification of outliers disturbing the homogeneity and lowering the number of data useful for the analysis. There will be two kinds of uncertainty: the just described—the inner uncertainty estimated either unrobustly as mean differences of results A and B or robustly by the parameters of the Ideal Gnostic Cycle. There also will be the outer uncertainty caused by differences of the mean AB value from the robust mean of all laboratories. The evaluation of uncertainty will be also done robustly in terms of entropies of the Ideal Gnostic Cycle: the quantification entropy f_I, the estimation entropy f_J, their difference (the residual entropy) and the relative entropy $(f_I - f_J)/(f_I + f_J)$ usable for comparison of difficulties of treatment of individual POPs and of laboratories. The inner uncertainties of laboratories are evaluated as penalties ordered by the total penalty in Tab. 23.2.

TABLE 23.2
Inner uncertainties evaluated by penalties for laboratories

Laboratory	Completeness	Censoring	Inner unc.	Suspect	Numerics	Tot. penalty
AB3	0.00	0.000	0.0389	0.0000	0	0.0389
AB8	0.00	0.000	0.1024	0.1250	0.1250	0.3524
AB12	0.25	0.000	0.0155	0.1875	0	0.4530
AB4	0.25	0.000	0.2781	0.1250	0	0.6531
AB6	0.25	0.125	0.0157	0.2500	0.1250	0.7657
AB7	0.25	0.125	0.0303	0.3125	0.0625	0.7803
AB9	0.00	0.625	0.0171	0.0625	0.1250	0.8296
AB13	0.25	0.375	0.0042	0.1875	0.1250	0.9417

The larger the inner uncertainty of individual laboratories, the stronger intensity of uncertainty of the contaminant. Judging that the measurement errors reveal the measurability of the matter, we can order the measurability of individual pollutants by robust means of residual errors of their Ideal Gnostic Cycle.

There are at least two ways of characterization of the uncertain or deterministic variability of a quantity:

1. To establish its range,
2. To quantify its uncertainty.

The contamination level of individual pollutants is in Tab. 23.3. It was obtained robustly by the function Gmean applied to the row of data table. Contamination range is the difference between the max() and min() of the row. The pollutants in Tab. 23.3 are ordered by the residual entropy (mean$(f_J - f_I)$) which is an important physical quantity as the upper bound of uncertainty once entered into the measured quantity. Its range is again the difference between its extreme values.

As seen in Tab. 23.3, the uncertainty of individual POPs substantially differs from 0.0010 (dibenzo[a,h]anthracene) through 0.5250 of naphtalene.

TABLE 23.3
Measurability of the POPs ordered by their residual entropy

POPs	Contam. value	Contam. range	Res. entropy	Entropy range
Dibenzo[a,h]anthracene	2.41	16.29	0.0002	0.0010
Acenaphtene	67.33	102.26	0.0006	0.0012
Pyrene	512.06	703.48	0.0012	0.0048
Fluoranthene	814.97	1154.16	0.0016	0.0076
Benzo[g,h,i]perylene	15.06	27.19	0.0017	0.0066
Acenaphthylene	8.61	12.95	0.0024	0.0110
Fluorene	102.88	152.20	0.0024	0.0163
Benzo[k]fluoranthene	18.86	30.23	0.0024	0.0072
Benzo[b]fluoranthene	58.22	114.63	0.0028	0.0103
Benzo[a]pyrene	13.11	25.25	0.0043	0.0124
Phenantrene	496.02	1031.16	0.0050	0.0256
Indeno[1,2,3-c,d]pyrene	13.55	21.22	0.0051	0.0207
Anthracene	32.46	60.25	0.0054	0.0171
Chrysene	79.23	183.91	0.0080	0.0502
Benzo[a]anthracene	73.01	773.83	0.0215	0.1389
Naphthalene	21.79	24.02	0.0914	0.5250

23.6 Homogeneity and Outliers

The homogeneity is one of main requirements to quality of a data sample. There are two causes of non-homogeneity, both are tested by the gnostic distribution functions:

1. Presence of an outlier or several outliers.
2. Splitting of the sample into two or more clusters

Both causes of non-homogeneity are manifested as additional local maxima of the probability density function, the outliers as individual local peaks and the separate cluster as a broad hill in parallel with the main cluster. The homogenization eliminates the outliers from the analysis and takes out the homogeneous main cluster by means of cluster analysis. In the case of these study is the homogeneity disturbed only by outliers. Looking for outliers requires computing of all 16 global distribution functions of all POPs done preliminarily by ignoring the censored data and followed by estimation of the censored data.

The following outliers were found in measurements of different POPs by individual laboratories: fluorene (laboratories 7, 12 and 1), phenantrene (lab. 7), fluoranthene (lab. 7), benzo[k]fluoranthene (lab. 12), dibenzo[a,h]anthracene (labs 4 and 12) and indeno[1,2,3-c,d]pyrene (labs 4, 7 and 12).

23.7 Estimation of Left-Censored Data

Before the final calculating of distribution functions, the problem of censored data must be solved. There were several difficulties of doing this.

Data weight: the weight of data item as well as its uncertainty depend on the item's value. The precision class of a measuring instrument is ordinarily established by its relative error (the ratio of absolute error divided by the instrument's range—its maximum measured value.) But when measuring a value close to zero, the absolute error is divided by a number decreasing to zero, the relative error rises.

Zero value: What is the value of lower measuring bound? It surely differs from the zero shown on the instrument's scale. It is a psychological problem, because everybody will declare the "value below the lower bound of measurement" differently. When measuring a value of sample's data, one must have the estimated lower bound LB of the homogeneous sample or of its homogenized sub-sample. Probability of data item less then LB is very small, LB can be taken as "measurement zero" and as the lower bound of the left-censored value. Its upper bound is the least value of the smallest uncensored sample's values. Anywhere between these bounds should be the censored value placed.

Distribution problem: The goal of analysis is to find data forming a homogeneous sample. Therefore it is logical to accept such left-censored data which will satisfy the requirement of homogeneity with the uncensored data.

Using these rules, the following values of left-censored data (LCDs were found: benzo[a]anthracene (laboratory 9), benzo[k]fluoranthene (lab. 12), benzo[a]pyrene (lab. 9), benzo[g,h,i]perylene (labs 9 and 13), dibenzo[a,h] anthracene (labs 6, 7, 9 and 13) and indeno[1,2,3-c,d]pyrene (laboratories 12 and 13).

In four upper cases the estimate of the left-censored data was successfully placed in a halve of the interval between the lower bound of the data domain LB and between the least from used data. The completed sample appeared to be homogeneous.

The last two cases were more difficult because there is evidence for problems in measuring these POPs. There were only two measurements of dibenzo[a,h]anthracene (4.8145 and 5.00) acceptable as actual measurement, but obvious outliers 16.4924 and 0.200 and four values declared as left-censored. The estimate of the LB (when considering only two more or less reliable measurements) were zero. The four estimates of left-censored data were put in the interval (0, 4.81) as approximately uniformly distributed. The resulting distribution function of six data was homogeneous. In the case of indeno[1,2,3-c,d]pyrene another problem existed: too narrow range of four actually measured data: 20.6282, 21.0, 22.9120 and 20.7846. The lower bound LB appeared to be close to the least of the measured data (20.61742) leaving thus only a very narrow interval for including the left-censored data. They were therefore chosen as confirming the measured value 20.6282. The resulting distribution was homogeneous.

The distribution functions of dibenzo[a,h]anthracene are shown as an example from 16 distributions of POPs in Fig. 23.1. All the distributions are homogeneous excluding outliers and included estimated values of left-censored data and estimated values of missing data (NA). All distribution functions differ from normal (Gaussian) course not only by their forms but also by their final data domains.

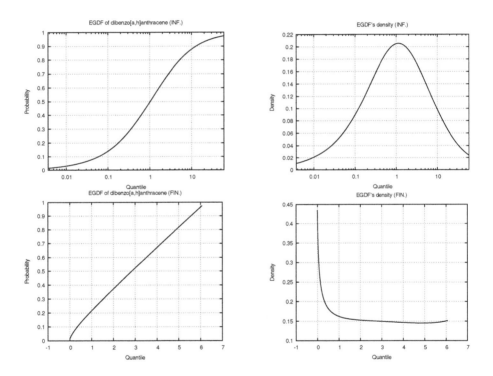

FIGURE 23.1
Homogeneous probability distribution functions of dibenzo[a,h]anthracene

23.8 Data Certification and Interval Analysis

Operations of automatic exploratory analysis and data certification are run by using the corresponding functions. What follows is a selection of outputs of these functions applied to phenantrene.

The data sample is multiplicative and homogeneous. Its recommended metric is 'E'. The data are to be considered homoscedastic. The effect of the data trend is 19.66%. The effect of the curvature is −5.41%. The mean absolute probability error of the distribution is 0.0432. The distribution functions over the natural (finite) and transformed (infinite) data support is in Fig. 23.2.

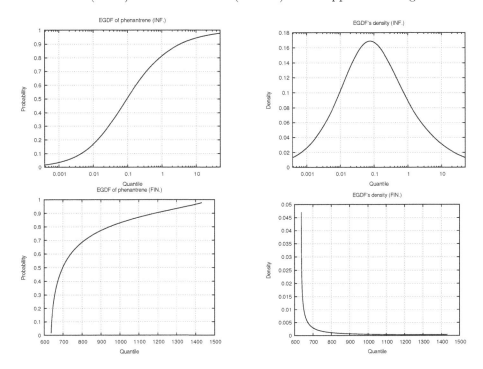

FIGURE 23.2
Homogeneous probability distribution functions of phenantrene

The most significant values of the autocorrelation function, main parameters of distributions, and bounds of interval analysis are summarized in Tables 23.4–7

This table enables the observed and estimated data of *phenantrene* to be classified:

The data sample is homogeneous, but its least value is close to the lower bound of homogeneity. The class numbers are denoted so that the interval

TABLE 23.4
Values of autocorrelation functions of phenantrene

Lag	Cor. coeff.	P. val.	Deg. of Fr.
Lag1	0.021	0.974	4
Lag2	0.463	0.382	3
Lag3	0.544	0.366	2
Lag4	0.139	0.897	1

TABLE 23.5
Main parameters of distribution function of phenantrene

Variable	Value	Name
Scaleopt	0.242	Scale parm.
Lowb	6.1e-21	Lower domain bound
Uppb	1780.4	Upper domain bound
InfQ	0.731	Information quality
MAFE	7.2e-05	Mean abs. fitting error
MXFE	0.00015	Max abs. fitting error
MAPE	0.0432	Mean abs. probability error
MXPE	0.1044	Max. abs. probability error
Kind	"G"	Kind of distribution
RResE	0.0067	Relative residual entropy

below the lower bound LB of the domain is zero. The data number 6, 4 and 7 appear to be below the lower typicality and they are classified as lower atypical. Only data item 5 falls into the upper interval of tolerance. The data No.2 and 1 are over the upper bound of typicality. Data item 3 lies close to the upper homogeneity bound. The data spread cannot be classified as too

TABLE 23.6
Classification bounds of phenantrene

Bound	Value	Cumul.Prob.	Name
LB	82.56	0.000	Lower domain bound
LHB	117.7	0.055	Lower homog. bound
ZL	468.8	0.458	Lower typicality b.
Z0L	478.1	0.468	Lower tolerance b.
Z0	549.1	0.542	Location parameter
Z0U	651.3	0.649	Upper tolerance b.
ZU	634.3	0.624	Upper typicality b.
UHB	1148.9	0.949	Upper homog. bound
UB	1263.7	1.000	Upper domain bound

TABLE 23.7
Classification bounds of phenantrene data

No.	Value	Probab.	Class
6	117.75	0.0553	3
4	326.30	0.3000	3
7	364.44	0.3425	3
5	617.48	0.6089	5
2	654.83	0.6429	7
1	727.03	0.7034	7
3	1148.91	0.9490	8

strong because only one of six data items can be considered as typical. Probability of 1/6 is 0.167 and the length of probability (ZU, ZL) of typical values $(0.624, 0.458)$ (Tab. 23.6 is practically the same. Recall the way of determination of interval bounds by interval analysis which makes use of hidden features of data and not some subjective assumptions.

Probability distributions of individual pollutants differ, therefore the ratios of bounds may be of interest (Tab. 23.8).

TABLE 23.8
Ratios of interval bounds of individual pollutants

Pollutant	$UB - LB$	UHB/LHB	UBT/LBT	UBt/LBt
Obj1	70.92	3.057	2.063	1.121
Obj2	15.20	7.171	1.032	1.007
Obj3	121.2	6.764	1.203	1.352
Obj4	7.976	1.050	1.025	1.012
Obj5	733.6	6.174	1.030	1.099
Obj6	63.29	9.313	4.104	1.483
Obj7	1361.	6.957	2.181	1.090
Obj8	825.1	7.196	2.163	1.082
Obj9	2824.	110.7	16.54	1.894
Obj10	255.9	Inf	4.398	1.434
Obj11	129.1	14.93	1.128	1.291
Obj12	17.62	1.453	1.372	1.083
Obj13	26.00	11.20	1.414	1.410
Obj14	12.22	1.675	1.046	1.014
Obj15	4.302	5.266	3.716	1.617
Obj16	2.557	1.108	1.002	1.000

23.9 Comparison of Laboratories

To compare laboratories by their outer errors we use means of available data pairs from Tab. 23.8 and divide them by means of all laboratories. The outer errors are evaluated by the ratio of the individual observed value divided by the mean of all laboratories.

TABLE 23.9
Individual multiplicative outer error of pollutants

Lab. 3	Lab. 4	Lab. 6	Lab. 7	Lab. 8	Lab. 9	Lab. 12	Lab. 13
1.001	0.467	1.570		1.098	1.147		1.010
1.244	1.747	1.567	1.155	0.985	1.510	0.244	0.581
1.387	1.026	1.195	1.782	1.366	1.433	0.264	0.512
1.023	1.011	1.005	1.391	0.989	0.974	0.215	0.425
1.558	1.643	1.480	2.596	0.737	1.395	0.266	0.824
1.855	1.436	0.981	2.079	0.703	1.808	0.223	0.601
1.654	1.018	1.241	1.410	1.210	1.130	0.238	0.863
1.596	1.027	1.184	1.357	1.362	1.169	0.222	0.887
1.354	13.66	1.634	1.739	0.811		0.195	0.295
1.676	2.317		1.275	1.175	0.597	0.186	0.816
2.110	0.395	1.271	2.061	1.500	1.571	0.141	1.380
1.217		0.965	1.019	1.000		0.116	
2.550	0.422	2.643	1.885	2.169		0.259	1.227
1.136	0.846	1.222	1.104	1.362		0.087	
0.981	3.361				1.019		0.041
1.498	1.125	1.525	1.270	1.663	1.509	0.123	

The large discrepancies between results delivered by different laboratories are worth of mentioning in the Tab. 23.9. We have already seen that measuring of organic pollutants is a difficult task, but the large systematic deviations of results of laboratory 12 should not be overlooked: all the results of this laboratory are below 26.6% of the joint mean. This laboratory generates results which are deep below the joint results of contamination. The quality control department of the laboratory should take measures to improve the results. Evaluation of individual laboratories is in Tab. 23.10. It is done by the ordering of laboratories by their order in evaluation of the inspection penalty, levels of outer errors and of the uncertainty measured by the mean residual entropies of the Ideal Gnostic Cycle.

TABLE 23.10
Ordering of laboratories

Lab. No.	Penalty	Out. Errs	Out. Unc.	Total
8	2	1	7	10
9	7	2	3	12
13	8	3	2	13
3	1	6	6	13
7	6	7	1	14
12	3	8	4	15
6	5	5	5	15
4	4	4	8	16

23.10 Conclusions

Advanced production control includes following steps:

Data inspection especially of respecting the FAIR principles.

Robust data means obtained by proper data aggregation.

Homogeneity test to decide on elimination of outliers or of outlying data sub-samples.

Interval analysis to classify the data with respect to membership in sub-intervals.

Data certification to summarize the results of analysis.

The exploratory analysis is done in gnostics automatically. A high level of automation is typical for all other gnostic functions. The decision which function is to be applied is on the analyst, but everything necessary for reaching the goal of analysis is prepared automatically and objectively because all decision making is based on data features. The optimality of results is reached by the theory which is supported by the laws of Nature.

24
Robust Correlation

Correlation methods are very popular because of their ability to quantify similarity between variables. The cross-correlation coefficient is defined as mean value of the products of centralized deviations of two variables divided by square root of the variables' variance. From the geometric point of view this coefficient represents a mean trigonometric cosine between a couple of inexactly measured linear processes. The uncertainty is entering into the coefficient several times:

1. Into the estimation of mean value of the first variable to obtain the centralized deviations.

2. Into the estimation of mean squares of the centered first variable to obtain its variance.

3. Into the estimation of mean value of the second variable to obtain the centralized deviations.

4. Into the estimation of mean value of the squares of centered second variable to obtain its variance.

5. In estimation of the mean product of centered variables.

Using the formula of the correlation coefficient presented in [23], we can see all the aggregating operations of making the summary values $\sum$ explicitly:

$$r = \frac{n\Sigma(xy) - (\Sigma(x))(\Sigma(y))}{\sqrt{(n(\Sigma(x^2)) - (\Sigma(x)^2)}\sqrt{(n(\Sigma((y^2)) - \Sigma((y)^2)}} \quad . \tag{24.1}$$

In the light of previous chapters we should mention that all seven operations are ordinarily done by using the aggregation operation applied to data which may introduce further errors. The problem of robustness thus makes estimation of correlations difficult.

There are three approaches to robust correlations in mathematical gnostics:

1. Via gnostic distribution functions.

2. Via regression models.

3. Via a combination of both with robust filtering.

All these techniques are robust and all of them have their advantages. They can thus represent a suitable solution under some special conditions.

24.1 Correlation via Distribution Functions

The production of different gnostic irrelevances can be met even in the simple formula of distances between two points. We already know, that the quantifying irrelevances are aggregated additively as hyperbolic sines unlike their angles (observed data). In an analogy fixed by the second gnostic axiom the estimating irrelevances are aggregated as trigonometric sines. Both types of irrelevances are robust estimates of data. In cases of weak uncertainty they both converge to the linear errors of observed data. It can be therefore expected that products of quantifying irrelevances and estimating irrelevances will play the role of generalization of the data products and that their normalized estimates of their means will be robust estimates of the correlations. This idea leads to following formulas:

$$\text{cor}_J = \sum_{i=1}^{L} \left(\frac{H_{a,i,J} H_{b,i,J}}{\sqrt{H_{a,i,J}^2 H_{b,i,J}^2}} \right) \tag{24.2}$$

and

$$\text{cor}_I = \sum_{i=1}^{L} \left(\frac{h_{a,i,I} h_{b,i,I}}{\sqrt{h_{a,i,I}^2 h_{b,i,I}^2}} \right) \tag{24.3}$$

where $H_{a,i,J}$ and $H_{a,i,I}$ are estimating irrelevances of a data sample $d_{a,1}, \ldots d_{a,L}$ obtained by the ratio $(h_J)/f_J)$ and $h_{b,i,J}$ and $h_{b,i,I}$ are estimating irrelevances of the data sample $d_{b,1}, \ldots d_{b,L}$ evaluated directly from data.

An interesting question is: how the irrelevances should be estimated? The linear formula connecting the estimating irrelevance with probability offers an advantageous answer: use a two-step solution:

1. Evaluate the homogeneous global distribution functions of both samples and estimate the irrelevances from probabilities.

2. Substitute the irrelevances into Equation (24.2) or (24.3) according to the proper geometry.

This method offers following advantages:

- Eliminates problems in interpretation of correlations caused by non-homogeneity of data.
- Enables making use of censored data.
- Unique setting of the scale parameter.

Correlations by Means of Regression 275

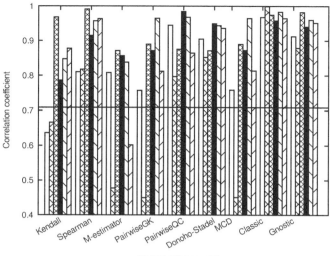

FIGURE 24.1
Comparison of statistical and gnostic methods of correlations

- Offers the opportunity to attach probabilities to arbitrary points.
- The proper data aggregation is respected.
- The finiteness of the data support is taken into account.

Figure 24.1 compares results of different methods of estimating the cross-correlations of contamination level of pairs of Czech and Moravian rivers.

24.2 Correlations by Means of Regression

There is an interesting alternative available to this approach in gnostics, application of robust regression models. Let x and y be centralized vectors of uncertain data of the same length, i.e. vectors of deviations from the mean. Simple regression models of their relations are

$$xc = y \quad yk = x \tag{24.4}$$

where

$$c = x^T/(x^T x) \quad k = y^T/(y^T y) \tag{24.5}$$

from which

$$\text{cor}(x, y) = \text{sign}(xy)\sqrt{ck} \tag{24.6}$$

is the standard correlation coefficient of the vectors. However, application of this formula can encounter problems in the case of strongly disturbed data. The method of IWLS (Iterated Weighted Least Squares) used for robust estimation of coefficients c and k does not find the least squares of errors of the Equation (24.4), but of the errors of these Equation weighted by weights $W1$ and $W2$ in cases of c and k, correspondingly. Components of these weighting vectors are optimized within the iteration process. Experience shows, that strong errors in x and y can cause large deviations of the product ck from its theoretic true undisturbed value 1. The outcome is to apply formula

$$\mathrm{cor}(x,y) = (xW^2 y)/\sqrt{(x^T W^2 x)(y^T W^2 y)} \qquad (24.7)$$

where W is the "best" weighting vector from $W1$ and $W2$ obtained as $\sqrt{W1 \cdot W2}$.

Following advantages are offered by this formula:

1. Robustly estimated regression coefficients c and k with the weights W will provide the correlation coefficient with robustness.

2. Uncertainties of estimates of regression coefficients are evaluated statistically. This enables the correlation coefficient's significance to be standardly tested.

3. Calculations do not need distribution functions, data are processed directly.

Decision between using this method and the one based on probabilities depends on the available data: to treat the data including the censored items or under the condition of not being sure, that data are homogeneous, apply the probability approach. Otherwise the faster direct method based on regressions could be more suitable.

As was pointed out in the beginning of this chapter, the uncertainty enters the estimated correlations several times by means of the statistical operation of estimating mean values which is aggregating additively the data instead of their thermodynamically determined functions. This may result in increasing uncertainty of the estimates of correlations. A further improvement of the robustness of correlation methods can be achieved by using the right aggregation in estimates of mean values.

24.3 Correlation and Filtering

Uncertainties disturb the estimation of the correlations. Data filtering reduce at least partially the uncertainty. It is therefore interesting to consider the effect of filtering on the correlation coefficient. Preparing an example take the shares market data, columns 1 (*Open*) and 5 (*Volume*) of rows 1:50 of matrix *Prices*. To evaluate the correlation coefficient, standard procedure of R-project *cor.test* with parameter "p" will be used running the classical (Pearson's) version of the correlation coefficient. Four steps of the example will follow:

1. Evaluation of the correlation coefficient of the "raw" observed data sets.
2. Filtering of both data sets by means of robust regression method using all five columns of matrix *Prices*.
3. Evaluation of the correlation coefficient of the filtered data sets.
4. Comparison of results of step 1 with step 3.

Tab. 24.1 shows the following results: *Lowb*—the lower bound of the 95% confidence interval, *Uppb*—the upper bound of the confidence interval, *CC*—the estimated correlation coefficient and *p-value*.

TABLE 24.1
Effects of filtering on the correlation coefficient

Data	*Lowb*	*Uppb*	*CC*	*p-value*
Raw	0.5225	0.8182	0.6991	1.6e-8
Filtered	0.8200	0.9389	0.8943	2.2e-16

Results confirm favorable impact of the filtration on the correlation coefficient, its value substantially increased, the confidence interval became narrower and the probability of the failure of the statistical test reached a minute value.

24.4 Autocorrelations

Autocorrelation is a method of quantifying correlations between the value in one point of a data series on the values in other points. In application to a time series it is the correlation of the "last" (most recent) value on the "old" data values. In cases of series of real data the task of estimation of the autocorrelation coefficient is as difficult as in cases of cross-correlation because of the uncertainty entering the data. The autocorrelation function of a time series summarizes the autocorrelation of the recent value with the lagged values. One of interpretation of the autocorrelation function is in its connection with the "memory" of the process: it demonstrates the degree of "remembering" the past values and their impact on the recent value. Another connection may be seen in inertial features of the process, e.g. what time is necessary for stabilization of the process after a disturbance. The information aspects can be also seen in autocorrelation of a time series: which information is available in lagged values to be used e.g. for prediction.

An example can illustrate the ideas (Fig. 24.2).

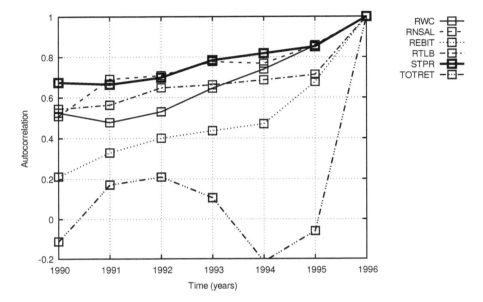

FIGURE 24.2
Autocorrelation of financial ratios of Chemical Industry of USA.

There is a large inertia in the basic financial ratios with an exception of the *TOTR*, in which recent value is impacted by the values four years old more than that of the last year. This also can reflect the situation of the market as a whole.

A comparison of the gnostic method using the irrelevances with several statistical ones has been done. Results are in Fig. 24.3.

The autocorrelations of daily temperatures of dataset *airq* were estimated by several methods available in R: classical Pearson's *OLS* along with methods of robust statistics *Huber*, *Hampel*, *Bisquare* and *Gnostic*. Evaluation of methods was performed by percentages of five classes of the p-values of the results. The results of gnostic method were the best in all classes.

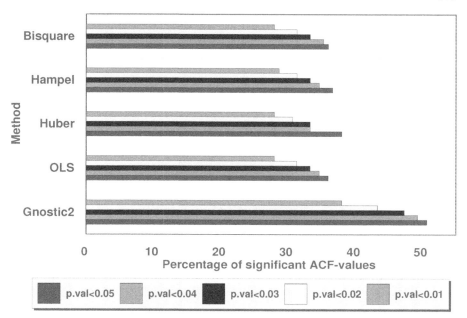

FIGURE 24.3
Comparison of several methods of autocorrelations

24.5 Conclusions

Robust correlations and autocorrelations represent useful instruments for getting insight into relations between vectors of uncertain data. Two methods were developed in mathematical gnostics for estimation of correlation coefficients, one using the gnostic distribution functions and the other based on the robust regression models. Two methods of data filtering are made available for improvement of the correlation analysis, one using the robust regression models and the other making use of the proper aggregation of the data characteristics. Although useful, the application of the correlation analysis is limited to studies of nearly linear processes.

25

General Relations

The notion of *correlation* is a special case of the more general word *relation*. It relates to a special situation of relation between two nearly linear processes or states quantified under uncertainties which (as shown in mathematical gnostics) are relatively small. This limitation makes them unsuitable for application to non-linear cases. Moreover, correlations are based on the first and second statistical moments which are not robust. However, processes or states changing in the mean in a non-linear way and under impact of strong uncertainties exist and their mutual relations are of great interest. Therefore the a more general case of relations deserves attention.

Let us consider a $N \times 2$ matrix M of strongly uncertain data. Names of M's rows are numeric $x = x_1, \ldots, x_N$. Several relations can be of interest:

1. $\{x, M[,1]\}$,
2. $\{x, M[,2]\}$,
3. $\{M[,1], M[,2]\}$.

Uncertainties in M's data make it necessary to require robust treatment of the relations. Tasks 1 and 2 can be also called "robust curve fitting" while task 3 is simply a relation.

Brackets $\{\,.\,\}$ instead of $(\,.\,)$ are used to emphasize that the relations are not necessarily functions $M[,K](x)$ neither $x(M[,K])$. (However, the first term will be called *argument* (A) and the other *value* (V), like in the case of a function). A general relation will be written as $\{A, V\}$). The rows of the matrix M can be ordered to get order of values of a numeric vector R which may be - but not necessarily—a column of this or other matrix. The ordered relation will be then written as $\{A, V\}|o(R)$. The ordering can reveal interesting features of data. So, e.g., ordering $\{A, V\}|o(V)$ enables the probability distribution of $V(A|o(V))$ to be estimated. Another example is $\{A, V\}|o(A)$ presenting the time evolution of V when time moments A of observation were not properly time-ordered. It is obvious that row-ordering of a multi-column matrix makes the number of relations based on this matrix large. However, the spectrum of such relations is further extended when the degree of required robustness controlled by the scale parameter is taken in account. Two classes of robustness are to be considered:

- Subjectively chosen degree of robustness.

DOI: 10.1201/9780429441196-25

- The unique (optimal) degree of robustness maximizing the resulting information.

There are good reasons to consider both classes: In dependence on the degree of robustness, the "smoothed" curve of A or V can have one or more extreme values resulting in changing directions of dependence of V on A in a point called *critical*. The critical points can be the most interesting results of analysis, but there is no guarantee that they can be found by using the optimal degree of robustness. The robust estimation of relations is considered below. Many unordered relations $\{A, V\}$ along with the ordered ones are used in *Advanced Relation Analysis*.

There are following goals of the relation analysis:

- To investigate matrices of uncertain data robustly by yielding the maximum information.

- To visualize the relations defined by a matrix.

- To provide an insight into multidimensional processes and/or states of objects.

- To identify critical points—values of variables where the course or behavior of variables significantly change.

- To enhance the expertise and support the control of processes and objects.

25.1 Relations Considered in Mathematical Gnostics

Following classes of relations will be considered in mathematical gnostics: Given one-dimensional sets $A(1),: A(2),: \ldots : A(m)$ and V of measured/observed numeric data.

1. A *bi-relation*: a rule $(VA(x))$ that maps an element of $A(x)$ uniquely to an element of V.

2. A *tri-relation*: a rule $(VA(1)|A(2))$ that maps an element of $A(1)$ and an element of $A(2)$ uniquely to an element of V under the impact of parameter $A(2)$.

3. A *multi-relation*: a set of tri-relations $(VA(x)|A(y))$ where x and y are from $1, 2, \ldots, m$.

4. An *auto-relation*: a set of bi-relations (VdV) that uniquely maps delayed elements of V to the last element of V.

These relations are not necessarily functions.

A multidimensional numeric matrix of R rows and C columns defines a set of relations:

- The i-th column can play a role of an argument $A(i)$ as well as of a value $V(i)$ of a relation. Relation of two columns is $V(i)A(j)$ $(i, j = 1\ldots, C)$. Excluding cases $i = j$ one can have $C^2 - C$ bi-relations between columns.

- Relations are not necessarily invertible.

- Denote $OrdO$ the ("original") sequence $1 : \ldots, : R$ of rows. Denote $OrdCk$ the sequence of rows ordered by values $V(k)$ of a column $C(k)$. Every pair $A(n), V(n)$ taken from the same row can thus be ordered in $C + 1$ ways to define tri-relations $V(n)A(n)|OrdCk$.

To demonstrate these notions, two examples from the medicine can be used: the interdependence of cytokines and their dependence on the neutrophiles. Cytokines are a broad category of small proteins that are important in cell signaling, specifically in host responses to infection, immune responses, inflammation, trauma, sepsis, cancer, and reproduction. Data used for examples were obtained by measuring on 15 patients of university hospital in Ostrava (25.1 and 25.2). Interdependencies of cytokines *VEGF*, *SDF*, *PDGF* and *VEDF* and their dependence on *Neutrophiles* are shown graphically in the figures.

There are three parts in the Fig. 25.1, call them α, β and γ. The sub-graphs α and β in Fig. 25.1 depict the bi-relations *VEDF Neutrophiles* and *SDF Neutrophiles*. The sub-graph γ shows the tri-relation $SVDVEDF|Neutrophil$ where the variable *Neutrophil* plays the role of a parameter. The dependent variables in α and β are depicted by heavy solid lines. The thin solid lines show the effects of sub-division of the data into odd and even. The small squares and spheres show values of individual data of both kinds. The squares of rising size show the values of the argument *neutrophiles*. The role of the argument in sub-graph γ is played by the variable *VEDF* which differs from the values of parameter *neutrophiles* under impact of which the course of the bi-relation changes. The size of large quadrates is rising in dependence on values of *Neutrophiles* from zero to maximum and the curve in γ passes five differing sections:

- Falling both *SVD* and *VEDF* from the starting point to the point $dVEDF/dSVD = 0$.

- Falling *SVD* and rising *VEDF* from the point No.1 to the point $dSVD/dVEDF = 0$.

- Rising of both *SVD* and *VEDF* from the point No.2 to the next point $dSVD/dVEDF = 0$.

- Falling *SVD* and rising *VEDF* from the point No.3 to the point $dVEDF/dSVD = 0$.

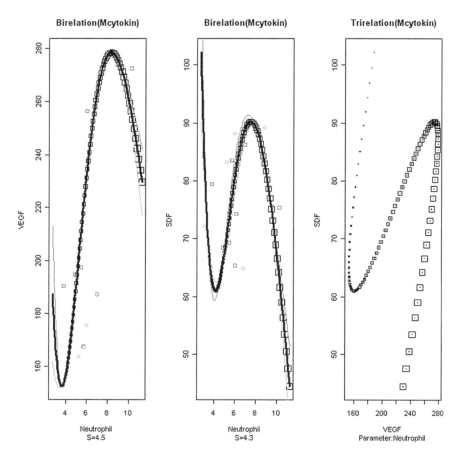

FIGURE 25.1
Examples of bi- and tri-relations in cytokines

- Falling both SVD and $VEDF$ from the point No.4 to the point of $SVD = 0$.

Such a complex interaction of cytokines is seen again in 25.2, where the role of the argument is played by the variable $PDGF$ instead of $VEDF$. The order of falling and rising phases of the process are here reversed.

The following three comments are due to these examples:

1. The information of splitting the interdependence of a multidimensional process in limited sections where the dependence can reverse is important not only for understanding of the process but especially for the process control. Imagine a physician aiming to improve a medical parameter of a patient by a drug and not being

Relations Considered in Mathematical Gnostics   285

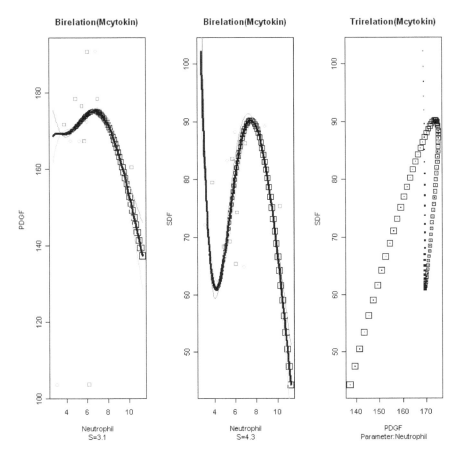

FIGURE 25.2
Examples of bi- and tri-relations in analysis of cytokines with critical points

sure if the drug will act on the parameter positively or negatively because this depends on the actual values of several parameters determining if the actual working point lies on rising or falling section of the process.

2. The examples show the unsuitability of the correlation analysis for such complex applications. Not only values of the correlations change locally, but their sign as well. Moreover, one single number (the value of the correlation coefficient) cannot characterize the complex dynamics of the process.

3. The points where the dependence reverses deserve a name. They will be called *the critical points*.

286   *General Relations*

The critical points can be identified numerically, because the equations of curves are known.

25.2 Robust Curve Fitting

The frequently recommended way of visualization of a binary relation is the application of the histogram: the values of a dependent value are depicted as series of discrete points. Such a figure does not represent the course of the variable sufficiently if the data are uncertain because the isolated points are placed chaotically. It is difficult to see the form of actual course of the process, to identify outlying data and local extreme values. Problems of smoothing by analytic functions was already discussed and smoothing by gnostic kernels recommended. There are two classes of problems remaining to be dealt with in connection with the curve fitting: the scale parameter and ordering of variables. They will be discussed during the discussion of examples of applications.

The value of the scale parameter S plays an important role in fitting of curves, because it decides on the flexibility of the fit: the smaller S, the more variable the curvature of the fit, the larger the number of its local minima and maxima. And opposite is also true: the larger S, the closer the fit to a linear function. There are two versions of the scale parameter, *freely set S* and *optimum S*. An example of a freely set scale parameter is in Fig. 25.3.

The freely chosen value 6 of the scale parameter results in fit of the variable ROA (Return on Assets) of IBM which has three minima and three maxima (the order of the ROA is original, numbers of year quarters). The medium of three smooth curves is the mean of odd and even fit of data, the small spheres denoting the odd and triangles the even observed data values. The distance between the odd and even fit depends on the value of the scale S. A fit obtained by means of freely set S will be called the *local fit*. The value of S along with the quantity $I = 0.6952$ (the relative information of the curve) are placed below the graph.

The optimization of the S is reached by its value 89.92 when the odd and even models coincide as shown in Fig. 25.4. The optimization is performed automatically. The fit obtained by optimum S will be called the *global fit*. The value of the relative information of the curve increased to 0.706.

25.3 The Experimental Mathematics

It might seem that once we have the possibility of automatic determination of the scale parameter, we do not need the chance to choose for the S freely an

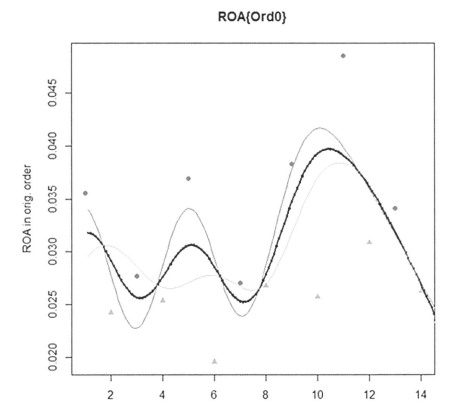

FIGURE 25.3
The local fit of *ROA* with freely set scale parameter (6)

arbitrary value. The problem is that the automated optimum—although existing always—is not always the best choice. Let us have an example (Fig. 25.5).

When one plays with the scale parameter, he may be surprised by the case depicted in Fig. 25.5: the odd and even model coincided at value $S = 30$ with an acceptable value of $I = 0.6402$. A question appears: Is this interpretation more plausible than the global interpretation? Answering this question would require further consideration. It reveals the nature of the relation analysis which can be called the *Experimental Mathematics*. Actually, the mathematics tends to precise and unique notions. Its style is "definition-theorem-proof." Reality of mathematical notions is ordinarily out of scope. Experiments ordinarily involve consideration of real things. But this is the aim of mathematical

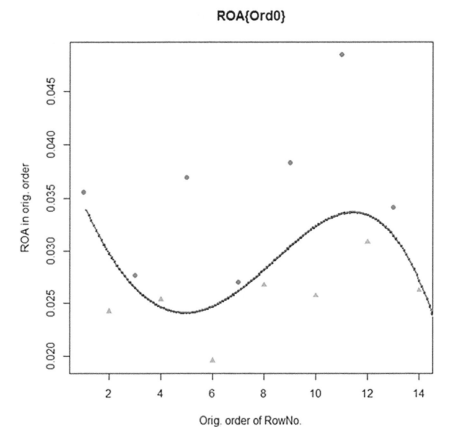

FIGURE 25.4
The global fit of ROA with optimum S

gnostics generally and of relation analysis specifically: to provide an analyst with mathematical instruments which allow reaching conclusions on facts of nature and life. The conclusions may be worth of the effort for the analyst, who plays with scale parameters, with ordering of variables, with combinations of variables and creative analysis of their behavior. Such an activity justifies the name of experimental. The instruments used and formulation of conclusions are mathematical.

The experiment can involve random decision making as well as some systematic actions. The latter way is demonstrated in Fig. 25.6 where the scale parameter was chosen at the initial case as 179.84 (double of the optimum value) and was followed by using the halves of the foregoing value. The scale

The Experimental Mathematics 289

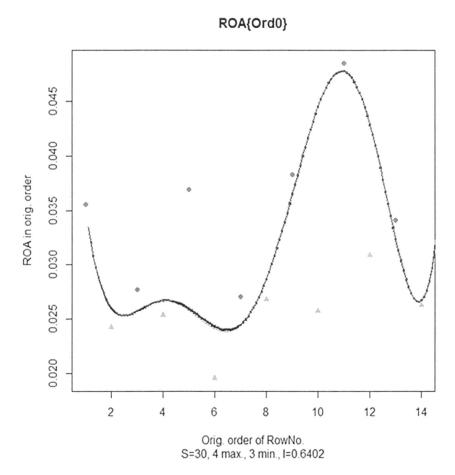

FIGURE 25.5
The local fit of ROA with a sub-optimum S

parameters on the last picture were obtained as $S = S0 \times SR$, where $S0$ was the optimal value maximizing the information. The SR was a multiplier taking gradually values $2, 1, 0.5, \ldots$. The impact of the scale parameter is obvious. The graphs show that it is not trivial to decide what is the best model. The best advice is to try different values of S and to look for symptoms of quality of the model. The important role is played by the number of local maxima and minima because they decide on appearance of the critical points. Another important indicator of graph's quality is the relative information I, which would prefer the third model with $S = 44.96$ which has the maximum value of I (0.7504).

FUNCTION *GFLS:* THE ROLE OF THE SCALE PARAMETER

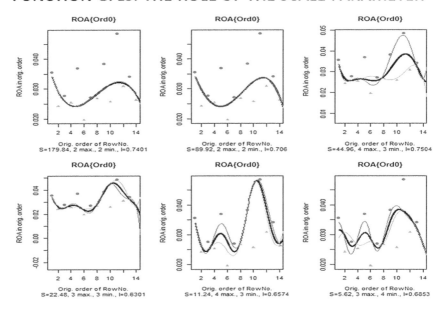

FIGURE 25.6
The local fits of ROA with decreasing S

25.4 Visualization of a Matrix

The robust curve fitting enables a multidimensional matrix to be visualized graphically. It can be done by using the local as well as global presentation. Both cases will demonstrate the financial data of the IBM. The local presentation is in Fig. 25.7 where the scale parameter is adjusted freely on values of halve of the optimum value.

The global presentation of the same matrix with optimum scale parameters is in Fig. 25.8.

What the graphs tell? Original row ordering is the time sequence: the graphs show the IBM's financial development. Increasing RTL (long-term debt) was enabling the working capital (RWC) to temporary rise. This resulted in acceleration of activity ($TATO$) and improving the profitability (ROA). Market was evaluating the process negatively by stock price which dropped more rapidly than was the ROA's increase as seen in P/E. The $TOTR$ did not follow the rise of ROA due to rapid fall of the stock price.

Some hints on interaction between variables can be obtained by exploring the impacts of ordering. The time changes are shown by $ROAOrd0$ because

Visualization of a Matrix

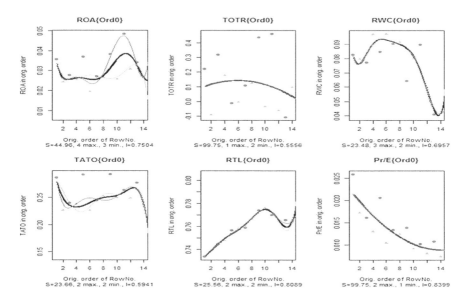

FIGURE 25.7
The local fits of financial ratios of IBM with $Sopt/2$

the original order is the time sequence. The least uncertainty is in the "auto-ordered" case of $TOTROrdTOTR$. Similarity of curves No.2 and 5 is a hint of strong connections between ROA and $TATO$. The reverse course of No.3 and 4 masks the actually positive impact of RWC on $TOTR$ due to rapid fall of stock price. (It may be, that market does not recognize the process properly).

A data matrix composed of numeric data of a process or of a system are useful as a database usable for the analysis. But a human being has only limited senses to understand fully some collections of numbers. The explanatory power of graphs helps to grasp not only data values but also their changes and relations to other data. A data matrix offers a rich choice of data information available by graphical presentation: global and local courses of all columns and of mutual dependencies of the columns ordered by their row numbers, by their values as well as by values of other columns. Such an album can serve the analyst in learning the reality which is reflected by data. An example of such an album is in Fig. 25.10.

The course of these graphs is unexpected. They do not satisfy the definition of a function because a function has just one value. However, all graphs of the album have at least two values over some intervals. These effects call for a more detailed analysis.

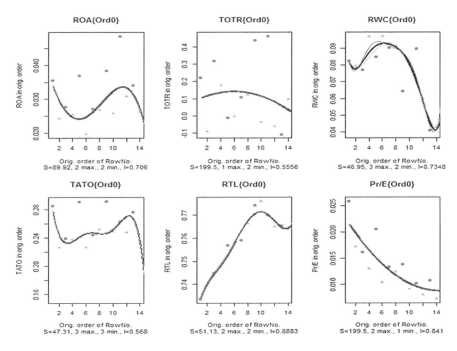

FIGURE 25.8
The global fits of financial ratios of IBM with S_{opt}

25.5 Critical Points

The "normal" behavior of a graph which is a function that respects the rule of the one-to-one mapping. It results in a curve which is rising, falling or constant depending on the increasing course of the argument. However, the graph of the album have sections over a limitedly rising argument followed after a point by falling section. Call this point the *critical point* as in the case of cytokins. The effect of critical points may be the most important result of analysis of the general relations. The nature of critical points can be demonstrated in more detail by the example of financial data of IBM (Fig. 25.11).

Neither the relation $ROA\,Ord0$ nor $RWC\,Ord0$ is monotonous, there are several extreme (critical) points, where the direction of changes reverses. Seven closed intervals are defined by these points. These are important for decision making.

The role of the working capital in maintaining the liquidity of operative debts of a business is well-known. However, keeping a sufficient part of capital for the application as RWC costs money and it can happen that the losses exceed the profit. The combinations of impact of changes in ROA and RWC lead

Critical Points

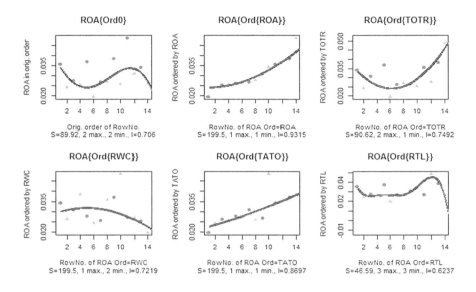

FIGURE 25.9
The impact of ordering on global fits of financial ratios of IBM with S_{opt}

to subdividing the relation $RWC(ROA)$ into five sections with different signs of the derivative $dROA/dRWC$ where only two subsections are acceptable. The knowledge of the critical points is important from the point of view of the financial management because the level of the RWC is under the control of the management unlike some other financial ratios determined by market.

TABLE 25.1
Critical points in financial ratios of IBM

Row No.	Argument FA	Value FV
Start	0.0816	0.0339
9	0.0777	0.0302
39	0.0912	0.0241
52	0.0930	0.0249
104	0.0658	0.0336
128	0.0411	0.0282
Finish	0.0560	0.0201

The availability of the information on relations of variables enables to reveal ways to rational control. This is demonstrated by the Fig. 25.12.

As shown in the first graph, the development of the ROA was steadily rising over the whole time of the observation of the process. The other relations show

294   *General Relations*

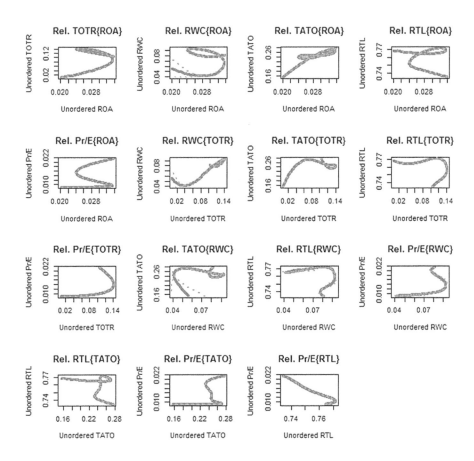

FIGURE 25.10
The global fits of financial ratios of IBM with *Sopt*

to the analyst what courses of other variable were necessary to maintain this behavior of the *ROA*: slight increase of the *RWC* would be insufficient, it was necessary to increase both *RTL* and *TATO*. This experience can be used as an advice for the future operations.

25.5.1 Relations in Biology

The first example of biological relations was already considered in Fig. 25.1 in connection with the interdependence of the cytokines. The important role of the *neutrophil* was obvious from these graphs. But for more, see Fig. 25.13.

Critical Points 295

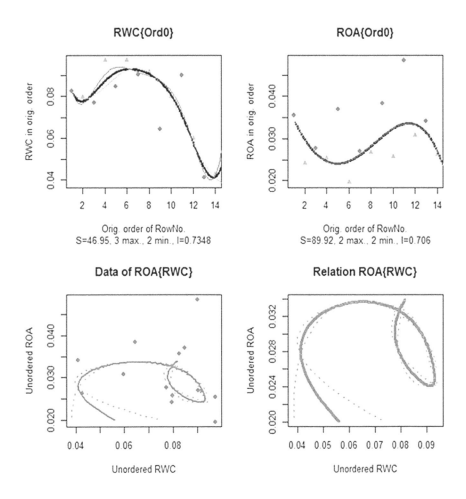

FIGURE 25.11
Critical points in relation ROA(RWC) of IBM

Relations between cytokines can be visualized by ordering by values of neutrophil. The complexity of the relations between cytokines is demonstrated in Fig. 25.14:

It is obvious from these figures that correlation coefficients cannot provide a true characterization of the interactions of variables (only the relation $VEGFNeutrophil$ is an exception).

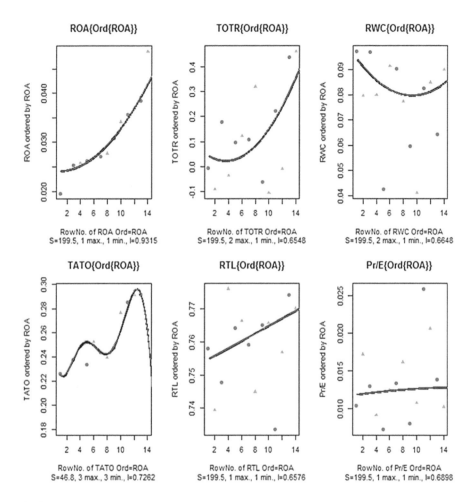

FIGURE 25.12
Relations as an experience to follow

25.5.2 Relations in Technology

Application of the technique of general relations to the famous data set stackloss demonstrates that relations between well measurable variables can be far from assumptions of correlation analysis (Fig. 25.15).

The non-linearity should not be ignored to interpret the data properly. A sharp peak/line break is worthy of interest in graph No.4.

Critical Points 297

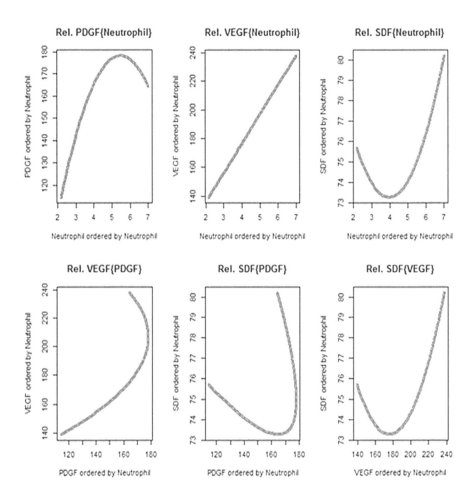

FIGURE 25.13
The role of neutrophil in dynamics of cytokines

25.5.3 Relations in Meteorology

Study of general relations may be useful even in case of natural processes which are not manageable but should be predicted like weather (Fig. 25.16).

The reversal of the relation $TemperatureWind$ may be surprising but its critical points are clearly defined by data (Tab. 25.2).

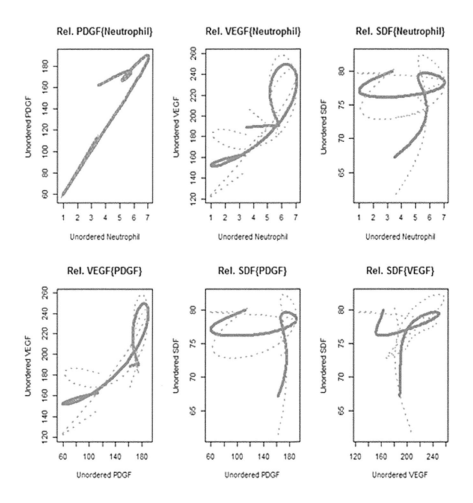

FIGURE 25.14
The complex interactions of cytokines

Graphs in Fig. 25.16 were obtained with the optimum value of the scale parameter. Figure 25.17 shows the same data when the scale parameter is chosen in a non-optimal way.

Using the scale parameter $S = 0.2$ enables some more flexible curves to be obtained. They offer a special interpretation of the observed process: the interdependence becomes nearly linear with similar slopes. Another set of critical points are obtained. Comparison of the figures helps to answer the question: why this way of analysis should be called *experimental mathematics*? The following reasons are the answers:

Critical Points

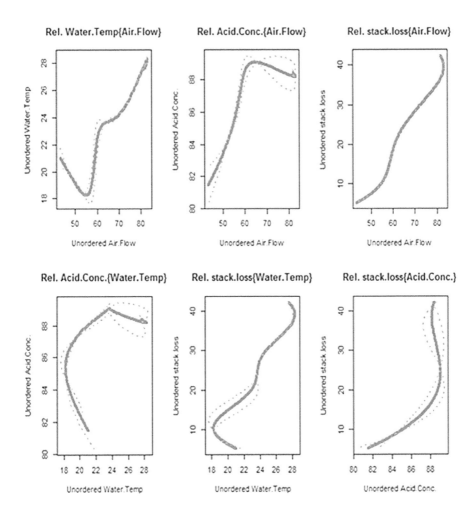

FIGURE 25.15
The non-linearity in technological data

1. Mathematics is a purely theoretical science. It generates such virtual worlds like the gnostic theory of individual data items and small samples.
2. Algorithms result from mathematical theories.

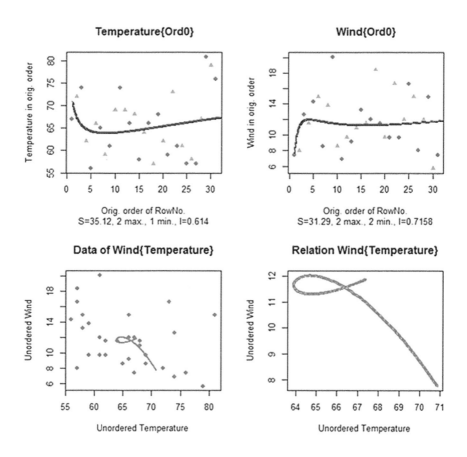

FIGURE 25.16
The interdependence of weather parameters with *Sopt*

3. To implement and use algorithms to solve problems of real world one needs real instruments for measuring and computers, elements of real world.

4. Complexity and uncertainty of real objects and processes do not allow to see everything in an unique and definite way.

5. Computer experiments using good algorithms help in looking for worthwhile information hidden in real data to enhance the knowledge of the natural processes and to manage them.

Critical Points

TABLE 25.2
Critical points of weather in New York

Point	Temperature	Wind
Start	70.85	7.76
31	64.69	12.00
70	63.90	11.63
149	64.76	11.29
Finish	67.35	11.85

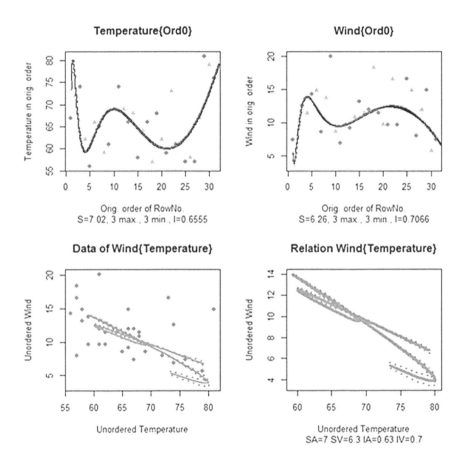

FIGURE 25.17
The interdependence of weather parameters with a non-optimal S

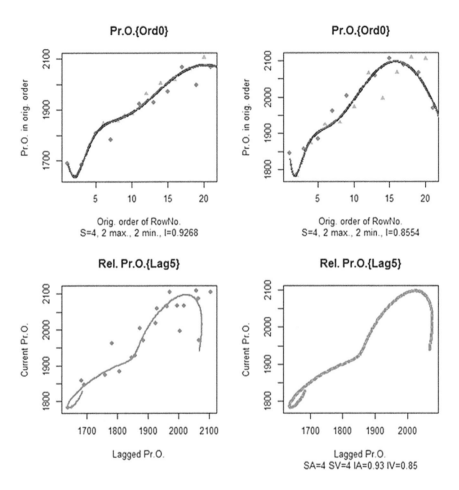

FIGURE 25.18
Auto-relations of the opening price of a share ($S = 4$)

25.5.4 Auto-Relations

Auto-relations are bi-relations of a recent part of a time-series with the same series but lagged one. It is a generalized notion of autocorrelation. The example shows the auto-relation of the opening price of a share (the first graph in Fig. 25.18) with its version lagged by six observation points (the second graph). The third graph summarizes the observed odd and even observed values and the fourth graph is the corresponding auto-relation. The example of time-series of opening price of a stock shows how inadequate would be the application of autocorrelation.

Critical Points

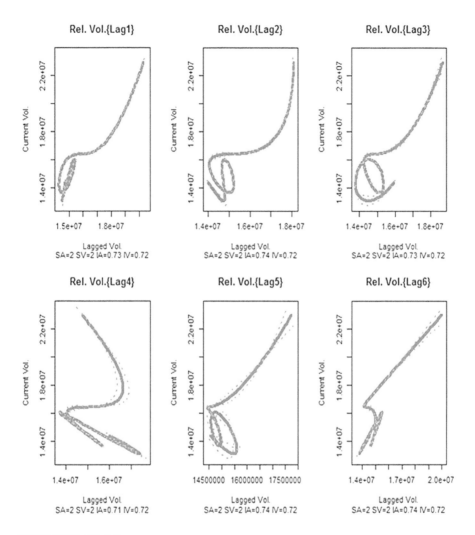

FIGURE 25.19
Auto-relations of the opening price of a share ($S = 2$)

The various figures in Fig. 25.18 were obtained with the scale parameter of the argument $SA = 4$ and of the value $SV = 4$ but lowering the scales to 2 enables more details of auto-relations to see in dependence on the lag (Fig. 25.19).

The auto-relations form changes depending on the lag.

The role of the scale parameter is important even in investigation of auto-relations. This can be illustrated by two versions of the auto-relations of the temperature in New York. The first (Fig. 25.20) was made with the scale parameter $S = 11$.

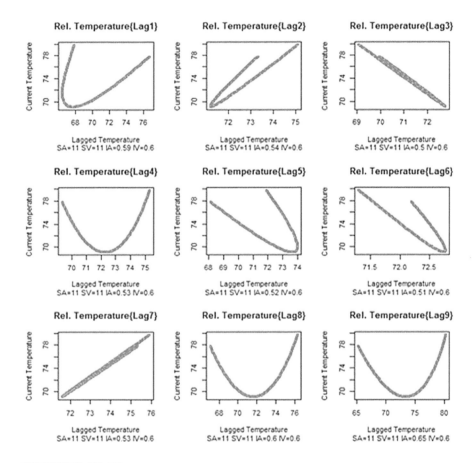

FIGURE 25.20
Auto-relations of temperature in New York ($S = 11$)

Decreasing the scale parameter to 10 changed the forms of curves significantly and revealed a series of new critical points.

This example shows the worth of experimentation with the relations: the change of scale parameter enables understanding of complex relations and helps to control them.

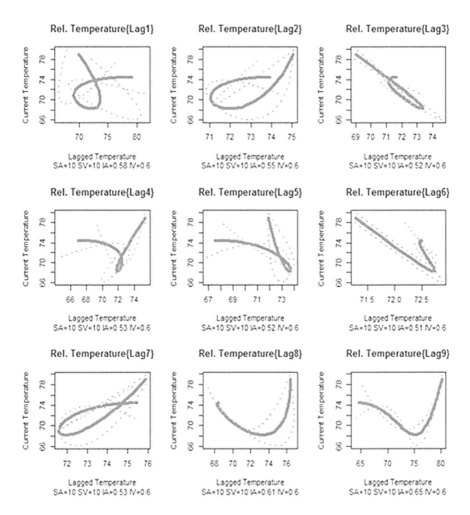

FIGURE 25.21
Auto-relations of temperature in New York ($S = 10$)

25.6 Conclusions

The application of gnostic kernels enables robust filtering of data from a complicated form including local maxima and minima. The form of the filtered variables is determined by the nature of processes and depends on the order of values and on scale parameters. Choosing of pairs or triads of variables enables to investigate the bi- or tri-relations in a manner more general than the cross- and autocorrelations which are applicable only for nearly linear interdependencies. However, variables of real processes frequently include local maxima and minima which result in appearance of reversal of dependencies called *critical points*. Observed data stored in the form of matrices include the information of existing critical points, but this information is hidden in data values. Revealing this information usable for understanding and control of the real processes is possible by gnostic algorithms when the variables, their ordering and scale parameters are properly chosen. The activity of combining the theoretical and real elements can be called the *experimental mathematics*. Its worth is in revealing or exposing the secrets of the Nature hidden in the data.

Bibliography

[1] Project FOKS, mathematical gnostics in WP3.

[2] Survey frontiers of finance. *The Economist*, (October 9):8–20, 1993.

[3] Project FOKS: project 1CE026P3 focus on key sources of environmental risks, March 2009–March 2012.

[4] 2FUN. Project 2-FUN (Full-chain and Uncertainty approaches for assessing health risks in future environmental scenarios), an integrated project of the European Union 6th framework programme (thematic priority 6.3 "Global change and ecosystems"), Feb 2007–Jan 2011. Contract n. FP6-2005-GLOBAL-4-036976.

[5] Abbott E. A. *The Annotated Flatland, A Romance of Many Dimensions*. Perseus Publishing, Cambridge, 2002.

[6] Baker J. A. and Christensen R. J. Discovering the future (The business of paradigms), 1989.

[7] Bloch A. *Murphy's Laws*. Price Stern Sloan, Inc., Los Angeles, 1993.

[8] Los C. A. A scientific view of economic data analysis. *Eastern Economic Journal*, XVII(1, Jan.–March):61–71, 1991.

[9] Markov V. A. *Cybernetic models of recognition*. Zinatne, Riga, 1976.

[10] Blažek B. Metoda explicitace zamlčených předpokladøu v ekologii (the method of concealed assumptions in ecology). *Bulletin ÚKE ČSAV*, 2: 1, 1977.

[11] Graham B. *The Intelligent Investor*. Harper & Row, New York, fourth, revised edition edition, 1973.

[12] Dostálek C., Kovanic P., Kufudaki O., and Málková L. Analysis of latent periods in decreasing learning curves. In *The 8th International Symposion of the INTERMOZG organization (Brain and Behavior)*, Liblice, 1984.

[13] Hrdinová D. and Kovanic P. Efficient solutions of the economic problems by means of gnostic algorithms. In *Proceedings of the conference "Application of Mathematic Models and Computers to the Control of National Economy"* pages 147–148. Economic University Bratislava, 1986.

[14] Kneale D. Into the void: What becomes of data sent back from space? Not a lot as a rule. *The Wall Street Journal*, (Jan. 12):1 and 33.

[15] Kovanicová D. and Kovanic P. High-tech for financial statement analysis. In *Proceedings of the 19th Annual Congress of the E. A. A.*, pages Abstracts, 328, full text 28, N. H. H. Bergen, Norway, 1996.

[16] Kovanicová D. and Kovanic P. Financial control of the growth rate of a firm. In *Treasures Hidden in Accountancy*, volume Part III. Polygon, Prague, 1997. in Czech.

[17] Kovanicová D. and Kovanic P. How to comprehend financial statements. In *Treasures Hidden in Accountancy*, volume Part I. Polygon, Prague, 1998. in Czech.

[18] Kovanicová D. and Kovanic P. Financial statement analysis. In *Treasures Hidden in Accountancy*, volume Part II. Polygon, Prague, 1999. in Czech.

[19] Luce R. D., Krantz D. H., Suppes P., and Tversky A. Representation, axiomatization, and invariance. In *Foundations of Measurement*, volume III. Academic Press, New York, 1990.

[20] Wessel D. Fickle forcasters. how three forcasters, after crash, revised economic predictions. *The Wall Street Journal*, (December 31):1 and 28, 1987.

[21] Kalman R. E. The problem of prejudices in scientific modeling. Final, written version of an invited lecture given on Sept. 4, 1986 at the European Econometric Meeting in Budapest, Hungary, with the title *Foundation Crisis in Econometrics within the Standard Statistical Paradigm*.

[22] Kochin N. E. *Vector Calculus and Principles of Tensor Calculus*. Publishing House of Academy of Sciences, Moscow, 1951.

[23] Triola M. F. *Elementary Statistics*. The Benjamin Cummings Publishing Company, Inc., Redwood City, California, 1989.

[24] Penman S. H. *Financial Statement Analysis & Security Valuation*. McGraw-Hill & Irwin, N. Y., 2001.

[25] Rocke D. H., Downs G.W., and Rocke A. J. Are robust estimators really necessary? *Technometrics*, 24(2):95–101, 1982.

[26] R. M. Heiberger and R. A. Becker. Design of an S function for robust regression using iteratively reweighted least squares. *J. Comp. Graph. Stat.*, 1:181–196, 1992.

[27] Berka K. *Measurements: its concepts, theories and problems*. D. Reidel Publishing Company, Dordrecht, Holland, 1983.

[28] Rashevski P. K. *Riemann's Geometry and Tensor Analysis.* GITTL, Moscow, 1953.

[29] P. Kovanic and M. B. Humber. *The Economics of Information (Mathematical Gnostics for Data Analysis).* https://www.math-gnostics.eu/books/, 2015.

[30] Fine T. L. *Theories of Probability; an Examination of Foundations.* Academic Press, New York and London, 1973.

[31] Mlodinow L. *The Drunkard's Walk.* New York: Random House, 2008.

[32] Coxeter H. S. M. *Introduction to Geometry.* Wiley, second edition, 1969. ISBN 9780471504580.

[33] Meloun M. and Militký J. *Kompendium statistického zpracování dat: metody a řešné úlohy (in Czech).* Academia, Praha, 2006. ISBN 80-200-1396-2.

[34] Stigler S. M. Do robust estimators work with real data. *Ann. Stat.*, 5 (6):1055–1098, 1977.

[35] Yaglom I. M. *A simple non-Euclidean geometry and its physical basis.* Springer Verlag, New York, 1979.

[36] MAGIC. Project MAGIC: Management of groundwater at industrially contaminated areas, 2005–2008. (5C028) EU-INTERREG III B CADSES.

[37] Taleb N. N. *The Black Swan.* Random House Inc., New York, 2010. 444 pp.

[38] NIST12. NIST WebBook Chemistry: 1,4-dichlorobutane. URL https://webbook.nist.gov/cgi/cbook.cgi?Name=1%2C4-dichlorobutane&Units=SI&cTP=on#Thermo-Phase. [cit 2020-05-06].

[39] NIST37. NIST WebBook Chemistry: Chloroform. URL https://webbook.nist.gov/cgi/cbook.cgi?Name=chloroform&Units=SI&cTP=on#Thermo-Phase. [cit 2020-05-06].

[40] Kovanic P. Guide to gnostic analysis of uncertain data.

[41] Kovanic P. Gnostic data treatment. *Chemický průmysl* (The Chemical Industry).

[42] Kovanic P. Minimum penalty estimate. *Kybernetika*, 8(5):367–383, 1972.

[43] Kovanic P. Generalized linear estimate of functions of random matrix arguments. *Kybernetika*, 10(4):303–316, 1974.

[44] Kovanic P. Gnostical theory of individual data. *Problems of Control and Information Theory*, 13(4):259–274, 1984.

[45] Kovanic P. Gnostical theory of small samples of real data. *Problems of Control and Information Theory*, 13(5):303–319, 1984.

[46] Kovanic P. On relations between information and physics. *Problems of Control and Information Theory*, 13(6):383–399, 1984.

[47] Kovanic P. A new theoretical and algorithmical basis for estimation, identification and information. In *The IX-th World Congress IFAC '84*, volume XI, pages 122–131, 1984. Preprints, IFAC Budapest.

[48] Kovanic P. Data analysis software based on the gnostical theory. In *International conference COMPSTAT '84*, Prague, 1984.

[49] Kovanic P. Introduction into the gnostic theory. In *Proceedings of the ROBUST conference '84*, pages 55–60. Faculty of Mathematics and Physics of the Charles University, Prague, 1984.

[50] Kovanic P. Gnostic algorithms for data treatment. In *Proceedings of the ROBUST conference '84*, pages 60–62. Faculty of Mathematics and Physics of the Charles University, Prague, 1984.

[51] Kovanic P. A program package for minicomputers for gnostic data treatment. In *Proceedings of the 5th conference on minicomputer technology*, pages 190–197. Dům techniky (House of Technology), Bratislava, 1984.

[52] Kovanic P. Gnostic monitor of processes GM1 for the SINCLAIR ZX-Spectrum computer, 1985.

[53] Kovanic P. Gnostic theory of data and its applications. In *Proceedings of the Colloquim "Selected Problems of Simulation"*, pages 12–28. Czechoslovak Scientific and Technological Society, Ostrava, 1985.

[54] Kovanic P. A new theoretical and algorithmical basis for estimation, identification and control. *Automatica*, 22(6):657–674, 1986.

[55] Kovanic P. A new theoretical basis for data analysis. *Automatizace*, 29(4):90–95, 1986.

[56] Kovanic P. Recent state of the development of the gnostic algorithms. In *Proceedings of the ROBUST conference '86*, pages 64–66. Faculty of Mathematics and Physics of the Charles University, Prague, 1986.

[57] Kovanic P. Verification of gnostic algorithms on the quality assessment problems. In *Proceedings of the conference "Application of Mathematic Models and Computers to the Control of National Economy"*, pages 243–244. Economic University Bratislava, 1986.

[58] Kovanic P. Gnostic monitor of probability GM3, project and program, 1986.

[59] Kovanic P. Simulation of empirical events. In *Proceedings of the 22-th spring colloquium "System Simulation and Mathematical Methods"*, pages 41–46. Czechoslovak Scientific and Technological Society, Ostrava, 1988.

[60] Kovanic P. Application of gnostics to identification of non-idealized systems. In *Proceedings of the KASIM 833 conference*, pages 49–57. Czechoslovak Scientific and Technological Society, Pilsen, 1988.

[61] Kovanic P. Gnostic diagnostics. In *Colloquium "Software of Diagnostic Systems of Nuclear Power Plants"*, ŠKODA Co., Pilsen, 1989.

[62] Kovanic P. *Gnostická teorie neurčitých dat. [Gnostic Theory of Uncertain Data]*. DrSc. diss., The Institute of Information Theory and Automation of Czechoslovak Academy of Sciences, Prague, 1990. URL http://www.math-gnostics.eu/download/MG12.pdf.

[63] Kovanic P. Applied gnostics. In *Conference of the IAEA "From Data to Model"*, Laxenburg, Austria, 1991.

[64] Kovanic P. Gnostic models. In *Proceedings of the MOSIS'92 colloquium, Acta MOSIS*, number 48, pages 62–65. Dům techniky (Technology House), 1992.

[65] Kovanic P. Robust estimates of distribution functions. In *Proceedings of the ROBUST conference '92*, Faculty of Mathematics and Physics of the Charles University, Prague, 1992.

[66] Kovanic P. A new model of surviving. In *Proceedings of the XIV-th Moravian colloquium "Selected Problems of Simulation Models"*, pages 31–34, MARQ, Ostrava, 1992.

[67] Kovanic P. *Financial Statement Analysis*, chapter 20, pages 259–290. Trizonia, Prague, 1993. (Upgrades of this book along with chapter 20 were republished by publishing house Polygon, Prague in 1994, 1995, 1996, 1997, 1998, 1999, 2000, 2001, 2002, and 2003).

[68] Kovanic P. Gnostical analysis of the cleanroom-like data, 1993. Interim report on research project ordered by the Digital Equipment Corporation (DIGITAL) in Vienna.

[69] Kovanic P. Gnostic filtration of acoustic signals, 1993. Development and research project ordered by a Czech Ministry.

[70] Kovanic P. Gnostical models of economics. In *International Conference MOSIS'93*, MARQ, Ostrava, 1993.

[71] Kovanic P. Intelligent load monitoring and forecasting system. In *SIMONE—International Workshop on Gas Distribution Systems*, pages 157–160, Prague, 1993.

[72] Kovanic P. Gnostical analysis of the cleanroom-like data, 1994. Final report on research project ordered by the Digital Equipment Corporation (DIGITAL) in Vienna.

[73] Kovanic P. Gnostic treatment of acoustic and thermovision signals, 1994. Development and research project ordered by a Czech Ministry.

[74] Kovanic P. Robust treatment of acoustic and thermovision signals, 1995. Development and research project ordered by a Czech Ministry.

[75] Kovanic P. Optimization problems of gnostics. In *Conference "Optimization-Based Computer-Aided Modelling and Design"*, Hague, April 2–4 (1991).

[76] Kovanic P. Gnostical modelling of uncertainty. In *MTNS '93— International Symposium on the Mathematical Theory of Networks and Systems*, Regensburg, Germany, August 2–6, 1993.

[77] Kovanic P. Gnostical approach to robust control. In *Joint British-Czechoslovak conference "Advanced Methods in Adaptive Control for Industrial Applications"*, Prague, May 14–16 (1990).

[78] Kovanic P. Smart matrices. In *IFAC Workshop on Mutual Impact of Computing Power and Control Theory*, pages 91–95, Prague, Sept. 1–2, 1992. Preprints MICC'92.

[79] Kovanic P. and Barack R. A. Gnostic analyzer GA5.2 for Windows, 1994. Commercial Program.

[80] Kovanic P. and Barack R. A. Robust survival model as an optimization problem. In *The 16-th IFIP Conf. on System Modelling and Optimization*, volume 2, pages 831–843, Compiègne, France, July 5–9, 1993.

[81] Kovanic P. and Humber M. B. A new paradigm for econometrics. In *The Third International Workshop on Artificial Intelligence in Economics and Management*, Portland, Oregon, U. S. A., August 25–27, 1993.

[82] Kovanic P. and col. Reliability of the presurized water reactors. In *Joint workshop of the Czechoslovak Academy of Sciences and ŠKODA Co., MODRA '87*, Modrá near Bratislava, 1987.

[83] Kovanic P. and Böhm J. Robust PID-control. In *IFAC Workshop on Mutual Impact of Computing Power and Control Theory*, pages 235–237, Prague, 1992. Preprints MICC'92.

[84] Kovanic P. and Novovičová J. A comparison of statistical and gnostical estimates of location parameters on real data. In *Proceedings of the ROBUST conference '86* pages 60–64.

[85] Kovanic P. and Pacovský J. Robust filtering and fault diagnosis by gnostical methods. In *Proceedings of the X-th World Congress IFAC 1987 on Automatic Control*, volume 3, pages 86–90, IFAC, Munich, 1987.

[86] Kovanic P. and Vlachý J. Gnostical analysis of international activities in physics. *Czech. J. Phys. B*, 36:71–76, 1986.

[87] Kovanic P. and Michajlov M. *Guide to the gnostic analyzer GA5.2 for Windows*, 1994.

[88] Kovanic P. and Kaprál R. Gnostic monitor of processes GM1. In *Proceedings of the conference "Mini- and Macrocomputers '86"*, volume Part III, pages 280–286. Czechoslovak Scientific and Technology Society, Prague, 1986.

[89] Kovanic P. and Prochazka V. Bmsc: The lifetime in control and experimental group.

[90] Kovanic P. and Žofková I. Medical experience with small data samples processing. In *Second European Congress on System Sciences*, Prague, Oct. 5–8, 1993.

[91] E. Parzen. On estimation of a probability density function and mode. *Ann. Math. Statist.*, 33(3):1065–1076, 1962. doi: 10.1214/aoms/1177704472.

[92] A. Perez. Mathematical theory of information. *Application of Mathematics*, 3(1):1–21, 1958.

[93] A. Perez. Mathematical theory of information. *Application of Mathematics*, 3(2):81–99, 1958.

[94] William H. Press, Brian P. Flannery, Saul A. Teukolsky, and William T. Vetterling. *Numerical Recipes—The Art of Scientific Computing*. Cambridge University Press, Cambridge, UK, 1986. ISBN 0-521-30811-9. Originally for FORTRAN and then rewritten for a C version, *Numerical Recipes in C*; also example books in Fortran, Pascal and C, making a total of five books with "Typeset in TEX" on the back of the title page.

[95] Hampel F. R., Ronchetti E. M., Rousseeuw P. J., and Stahel W. A. *Robust Statistics, The Approach Based on Influence Functions*. Wiley, New York, 1986.

[96] R language. *R: A Language and Environment for Statistical Computing*. R Foundation for Statistical Computing, Vienna, Austria, 2010. ISBN 3-900051-07-0.

[97] R manual. *An Introduction to R*. URL https://cran.r-project.org/manuals.html.

[98] Kuhn T. S. *The Structure of Scientific Revolutions*. The University of Chicago Press, third edition edition, 1969.

[99] Paukert T., Rubeška I., and Kovanic P. A new look at analytical data through the gnostical analyser. *The Analyst*, 118:145–148, Febr. 1993.

[100] B. N. Taylor and C. E. Kuyatt. Guidelines for evaluating and expressing the uncertainty of NIST measurement results. Technical report, NIST Technical Note 1297, 1994.

[101] Linnik Y. V. *The Least Squares Method and Basics of Observation Treatment*. Phizmatiz, Moscow, 1962.

[102] Pinta V. and Kovanic P. Příspěvek k metrologii pražských grošů Karla IV. (1346–1378). *Numismatické Listy*, LV 5/6:142–147, 2000. (A Contribution to the Metrology of Grossi Pragensis, in Czech).

[103] M. Černá, J. Kratěnová, K. Žejglicová, M. Brabec, M. Malý, J. Šmíd, Š. Crhová, R. Grabic, and J. Volf. Levels of PCDDs, PCDFs, and PCBs in the blood of the non-occupationally exposed residents living in the vicinity of a chemical plant in the Czech Republic. *Chemosphere*, 67(9): S238–S246, 2007. ISSN 0045-6535. doi: 10.1016/j.chemosphere.2006.05. 104. Halogenated Persistent Organic Pollutants Dioxin 2004.

[104] Baeyer H. Ch. von. *Maxwell's Demon*. Random House, N.Y., 1998.

[105] Helmholtz H. von. Zaehlen und Messen erkenntniss-theoretisch betrachtet. In *Philosophische Aufsaetze Eduard Zeller gewidmet*, pages 17–52. Leipzig, 1887.

[106] Dodge Y. *The Guinea Pig of Multiple Regression*, pages 91–117. Springer New York, New York, NY, 1996. ISBN 978-1-4612-2380-1. doi: 10.1007/978-1-4612-2380-1_7.

Index

'E'-metric, 258
'Q'-metric, 258
'R'-metric, 258

a posteriori data weights, 258
a priori
    model, 23
a priori data weights, 258
A.Perez's information, 65
Abelian
    group, 2
    group, 29
absolute variability, 184
accounting
    prudence, 148
additive, 1
    data, 2
    data, 29
additive aggregation, 48
additive data, 80
Additive Data Model, 11
additivity of probability, 67
advanced
    data analysis, 23
advanced comparison, 252
aggregation, 23, 33, 55
aggregation computing, 173
aggregation in mechanics, 56
agnostic, 31
agnosticism, 22
agnostics, xii
alcohol, 231
algorithms, xvii
alternative hypothesis, 95
amputation, 151
anti-information, 65
arsenic contamination, 152

asked price, 149
associativity, 2
astronomy, 50
auto-relations, 302

best mean, 185
bi-relation, 283
bid price, 149
blood pressure, 96
Boltzmann's entropy, 61
bounds of data domain, 24
bounds of data support, 24
bounds of membership, 125

cadmium, 136
censored data, 24, 143, 150
censoring, 143
Clausius, 8
Clausius' entropy, 61
clean matrix, 236
closure, 2
cluster, 24
cluster analysis, 83, 84
cobalt, 136
coins'
    degradation, 93
    purity, 92
    quality, 92
    weight, 92
collective error, 19
combined standard uncertainty, 182
commutative group, 24
commutativity, 2
comparability, 24
compared methods, 216
comparison advanced, 252
comparison by errors, 253

comparison by parameters, 254
comparison by the norm, 250
comparison evaluation, 219
comparison experience, 221
comparison of geometries, 216
comparison results, 220
comparisons, 181
concealed assumption, xvi
consistency, 181
contamination factors, 98
contamination rate, 98
continental contamination, 232
coordinate, 4
correlations by regression, 275
correlations in gnostics, 273
cost, 145
counting, 29
criteria NIST, 181
critical points, 285, 292
curvature, 24, 213

data, 1, 17, 29
   uncensored, 144
   classes of errors, 68
   analysis
      advanced, 23
   censoring, 143
   classes of errors, 68
   finiteness, 24
   interval, 143, 149
   left-censored, 143, 145
   lower atypical, 114
   normality, 119
   right-censored, 143
   speaking for themselves, 214
   survival, 147
   trimming, 23
   typical, 114
   uncensored, 143
   weight, 24
   right-censored, 147
data additivity, 258
data calibration, 159
data Cauchy, 177
data censoring, 258

data classification, 115, 167
data comparison, 218
data generation, 182
data heteroscedastic, 51
data homoscedastic, 51
data information, 112
data model, 32
data multiplicativity, 258
data NIST12, 176
data NIST12H, 177
data NIST37, 161, 167, 176
data objects, 218
data psychology, 153
data rnorm, 177
data stackloss, 85
Data structure, 82
data support, 6
data support bounds, 258
data swiss, 187
data TOTR IBM, 178
data upper atypical, 114
data violation, 202
datacratic role, 245
decision by means of critical points, 292
decision making in gnostics, 245
decision risk, xvi
defects of official comparison, 250
Design of EGDF, 87
different means, 196
different robustness, 49
dimension, 1
dimension of a matrix, 4
distance, 45
distribution
   functions, 24
distribution choice, 82
distributions tasks, 82
diversity of samples, 115
domain, 6
domain bounds, 81, 82
double numbers, 7
double robustness, 217

E-entropy, 208

Index

E-information, 208
E-space, 13
E-uncertainty, 208
econometrics, 21, 22
economical aspect, 112
EGDF, 80, 135
    application, 93
eigen decomposition, 5
ELDF, 80
ELDF applications, 107
empirical distribution, 86
entropies estimation, 171
entropy, xii, 25, 61
entropy ↔ information conversion, 64
entropy Boltzmann, 61
entropy Clausius, 61
entropy measures, 159
entropy Shannon, 61
environmental control, 50
error
    function, 209
error collective, 19
error individual, 19
estimate, 46
estimating
    distribution, 75
estimating data variability, 13
estimating global distribution
        function, 80
estimating local distribution
        function, 80
estimating relevance, 13
estimating space, 13
estimation, xii
    path, 35
ethanol engine, 154
Euclidean
    error, 210
Euclidean geometry, 8
Europe contamination, 99
event
    random, 18
expanded uncertainty, 183
experience 1, 140

experience 2, 141
experimental mathematics, 286
explicit model, 222
explicit regression, 205, 223
exploratory analysis, 257
extremal, 33
Extremistan, 70

fat, 232
feed-back, 223
filtered data, 258
filtering, 174, 235
    weight, 209
financial problems, 50
financial ratios, 122
fitting errors, 258
fuzzy set, 123

G-uncertainties, 37
Gaussian distribution, 120
Gedanken-experiment, 61
gedankenscience, 21, 22
general relations, 281
geodesics'
    gradients, 35
geometry, 7, 24, 46
    Euclidean, 46
    paradigm, 17
    Riemannian, 46
global
    distribution, 75
global distribution, 258
global fit, 286
global probability, 80
gnostic, 25
    criterion functions
        review, 208
    kernel, 74
    paradigm, 20
    regression
        robustness, 208
gnostic aggregation, 171
gnostic characteristics, 14
gnostic criterion functions, 205
gnostic cycle, xii

gnostic specifics, 217
gnostics
 consistency with statistics, 67
granit, 109
Grossi Pragensis, 91

healthy state, 121
Helmholtz H. von, 7
heteroscedascity, 3
heteroscedastic data, 51, 80
heteroscedasticity, 24, 199, 202
historical analysis, 95
historical coins, 91
homogeneity, 24, 99, 258
 definition, 126
homogeneous data, 83
homogenization, 83, 101, 102, 126, 192
homoscedascity, 3
homoscedastic data, 51
homoscedasticity, 24, 199, 202
hormon lh, 190
hypotheses testing, 83

ideal
 value, 46
 quantification, 29
Ideal Gnostic Cycle, 39
identity
 element, 2
IGC, 39
IGC—Ideal Gnostic Cycle, xii
impact of ordering, 290
impact varS, 203
implicit model, 222, 236
implicit regression, 224
improving matrices, 238
in-line monitoring, 200
indeterminate, 36
indeterminism, xi
individual data item, xi
individual error, 19
influence functions, 216
information, 25, 65
 historical, 91

Information in MG, 65
information perpetuum mobile, 40
information quality, 258
inlier, 24
inner
 robustness, 25, 211
input variable, 207
inspiration for gnostics, 6
inter-science
 isomorphism, 56
inter-science isomorphism, 57
interval
 data, 149
interval analysis, 84, 113, 114
interval data, 83
interval of lower atypical data, 114
interval of typical data, 114
interval of upper atypical data, 114
interval tolerance, 114
inverse
 element, 2
inversion, 5
IPMU, xv
isomorphism, 3, 56
isomorphism statistics-mechanics, 48

K-S point, 87
Kaplan-Meier method, 146
kernel data, 166
kernel matrix, 165
kernel triplication, 166
kernel estimate, 73
Kinds
 of
  distribution, 75
Kolmogorov-Smirnov test, 86

Least Squares Method, 56
left-censored data, 83, 150
 examples, 145
life-time, 147
LIFO, 145
limit of detection, 150
limitations of correlations, 273
Linnik's theorem, 18

Index   319

living place, 231
local
    distribution, 75
local distribution, 258
local probability, 79
local transport, 232
location parameter, 82, 115
Lorentz transformations, 58
lower atypical interval, 115
lower improbable interval, 115
lower membership bound, 115
lower tolerated interval, 115
lower typical interval, 115

marginal analysis, 84, 110
market
    price, 149
mathematical
    structure, 29
mathematical gnostics, xvi, 25
matrix, 1, 4
Maxwell's demon, 61, 63
mean error, 186
mean values, 173
measurement
    theory, 29
measurement errors, 254
measurement theory, xvi, 7
measurements
    low sensitivity, 145
    off the scale, 148
measuring, 29
    unit, 29
measuring methods, 109
Mediocristan, 70
membership
    problem, 24
membership function, 123
metric, 24
    tensor, 51
MG, xvi
minimum variance, 205
Minkowskian
    circle, 33
Minkowskian geometry, 8

missed scientific revolutions, 47
model, 15
    implicit, 224
    regression
        additive, 215
        multiplicative, 215
model explicit, 222
model implicit, 222
muliplicative
    data, 2
multiplicative, 1
    data, 29
Murphology, 16

negligible data, 24
New York ozon, 177
NIST12 homogeneity, 191
non-homogeneity, 23
non-smooth
    behavior, 23
non-statistical
    methods, 24
    paradigms, 19
normal range, 122
normality, 119
normalized
    aggregation, 75
null hypothesis, 95

observed
    data, 32
observed data, 11
OLS, 209
operation, 1
optimum path, 25
Ordinary Least Squares, 209
ordinary least squares, 70
Ordinarystan, 71
outer robustness, 25, 211
outlier, 23, 24
outliers clusters, 258

pair numbers, 7
paradigm, xvi, 15
    Galilean, 16
    gnostic, 20

non-statistical, 19
parts, 15
problems, 16
Ptolemaic, 16
risks, 16
statistics, 17
paradigm advantage, 15
paradigm limitation, 16
path
    integration, 45
point estimate, 24
point estimates, 109
political aspects, 112
pollutants, 96
POPS, 107
price decision, 157
probability, 258
    theories, 18
probability density, 6, 258
probability distribution, 24
probability global, 80
probability local, 79
product's quality, 149
project BONUS, 129
proper weights, 174
pseudo-inverse, 5

Q-entropy, 208
Q-information, 208
Q-irrelevance, 36
Q-regression, 209
Q-space, 13
Q-uncertainty, 208
QGDF, 135
quality
    of coins, 92
quality assessment, 50
quantification, xii
quantifying
    distribution, 75
quantifying data variability, 13
quantifying distribution, 85
quantifying irrelevance, 13
quantifying space, 13
quantile, 6

R-project, 215
radius
    vector, 39
random
    event, 18
randomness, 22, 25
rank, 5
rational control, 293
redescend function, 212
reference range, 122
reference value, 162
reference values, 50, 121
regression
    robustness, 206
regression filtering, 236
regression model, 19
relation, 281
relations classes, 282
relations in biology, 294
relations in finance, 286
relations in meteorology, 297
relations in technology, 296
relative
    frequency, 18
relative errors, 251
relative uncertainty, 186
relativistic physics, 8
requirements to quality of
    measurements, 248
residual entropy, 39, 40, 62, 171, 258
Riemann's geometry, 51
Riemannian
    metric, 210
Riemannian geometry, 8
right-censored data, 83, 150
rivers contamination, 136
robust curve fitting, 286
robustness, xv, 25
    regression, 206
robustness kinds, 50
robustness objective, 281
robustness subjective, 281
roles of relation analysis, 282
RRE, 186

Index

S-PLUS, 215
saturating function, 212
scalar, 1
scale parameter, 11, 75, 258
scale variability, 202
selling price, 145
separation of variabilities, 24
Shannon's entropy, 61
significance, 120
similarity, 24
singular value decomposition, 5
smoking impact, 231
society decline, 94
source
    of the entropy field, 208
standard
    deviation, 3
standard deviation, 182
standard domain, 81
statistical
    paradigm, 18
statistical comparison, 250
statistics, 17, 31
    successes, 20
STD, 3
strangeness, xi
sufficiently precise datum, 67
survival analysis, 83
swiss data, 187

tax expense, 148
taxes collection, 155
the lower bound LB, 81
the upper bound UB, 81
theory of measurement, 7
thermodynamic filtering, 237
thermodynamic weight, 49
thermodynamics, 8
three interpretation, 154
three mean values, 183
tolerance interval, 114
toxicity, 138
transformation, 56
trends, 182
tri-relation, 283

true values, 11
two entropy fields, 64
type I error, 95
type II error, 95
Types
    of
        distribution, 75
typical data, 114

unbiasedness, 70, 205
uncertainty, 2, 11, 25, 30
    paradigm, 17
uncertainty aggregation, 58
uncertainty classes, 68
uncertainty determination, 24
uncertainty in IBM data, 188
uncertainty in Swiss data, 187
uncertainty mapping, 58
uncertainty measures, 159
uncertainty unit, 184
underestimation, 145
undeterminate, 7
unknown limit, 150
upper atypical interval, 115
upper improbable interval, 115
upper membership bound, 115
upper tolerated interval, 115
upper typical interval, 115

variability, 2
variability information, 195
variability measuring, 183
variability weather, 190
variance, 3
variance controlled, 3
variance total, 3
variance uncertain, 3
variation principle, 33
varS parameter, 258
vector, 1
visualization of a matrix, 290

weak uncertainty, 67
WEDF
    interval data, 149
    left-censored data, 145

right-censored data, 149
weight
    variable, 46
weighted
    average, 46
weighting functions, 46
Why experimental mathematics?, 298